CONGRESSIONAL POLICIES, PRACTICES AND PROCEDURES

TOP 5 REPORTS AND TESTIMONIES IN AUGUST 2018

CONGRESSIONAL POLICIES, PRACTICES AND PROCEDURES

Additional books and e-books in this series can be found on Nova's website under the Series tab.

CONGRESSIONAL POLICIES, PRACTICES AND PROCEDURES

TOP 5 REPORTS AND TESTIMONIES IN AUGUST 2018

CHAD JOHNSTON

AND

WINIFRED PAYNE

EDITORS

snova

New York

NOTICE TO THE READER

Library of Congress Cataloging-in-Publication Data

ISBN: 978-1-53615-511-2

Published by Nova Science Publishers, Inc. † New York

CONTENTS

PREFACE

This book is a compilation of significant reports and testimonies issued in August 2018.

Chapter 1 - One of the Bureau's most important functions is to conduct a complete and accurate decennial census of the U.S. population. The decennial census is mandated by the Constitution and provides vital data for the nation. The Bureau plans to significantly change the methods and technology it uses to count the population with the 2020 Census, such as by offering an option for households to respond to the survey via the Internet. In preparation for the 2020 Census, the Bureau is conducting a test of all key systems and operations (referred to as the 2018 End-to-End Test), which began in August 2017 and runs through April 2019. GAO was asked to review the Bureau's IT readiness for the 2020 Census. This report (1) determines the Bureau's progress in developing and testing systems for the 2018 End-to-End Test and (2) describes the challenges and risks that the Bureau has faced in implementing and securing these systems. To do this, GAO reviewed key documentation, including plans for system development and testing, and outcomes of key IT milestone reviews and security assessments.

Chapter 2 - The Navy is planning to dismantle and dispose of CVN 65 after 51 years of service. In 2013, the estimated cost to complete the CVN 65 work as originally planned increased to well over $1 billion, leading the

Navy to consider different dismantlement and disposal options. The Senate Report accompanying a bill for the National Defense Authorization Act for Fiscal Year 2018 included a provision for GAO to review the Navy's plans for CVN 65. This report addresses (1) dismantlement and disposal options under consideration; (2) nuclear regulatory authority considerations; and (3) funding and reporting practices. GAO reviewed budget, cost, and schedule documentation, as well as applicable laws, regulations, executive orders, policies, and guidance. GAO interviewed officials from the Navy and commercial companies about the dismantlement and disposal options, and NRC and state agencies about regulatory considerations.

Chapter 3 - As of 2014, DHS estimated the total alien—a person who is not a citizen or national of the United States— population in the United States was about 27.1 million. Members of the alien population that have been arrested and convicted of crimes in the United States are referred to as criminal aliens. The costs associated with incarcerating criminal aliens are borne by the federal government, states, and localities. GAO was asked to update its March 2011 report on criminal alien statistics. This report addresses, among other things, the (1) number and nationality of incarcerated criminal aliens, (2) number of criminal alien arrests and convictions, (3) estimated costs associated with incarcerating criminal aliens, and (4) experiences of criminal aliens after incarceration in federal prison. GAO analyzed the most recent data available on criminal aliens, generally from fiscal years 2010 through 2016. Specifically, GAO analyzed data for criminal aliens incarcerated in federal prisons from fiscal years 2011 through 2016, and this group is the federal criminal alien population studied for this report. GAO also analyzed data for certain criminal aliens incarcerated in state prisons and local jails that received SCAAP funding from fiscal years 2010 through 2015, and this group is referred to as the state and local study population reviewed for this report. SCAAP data represents a portion of all criminal aliens incarcerated in state prisons and local jails. GAO used SCAAP data because there are no reliable data available on all criminal aliens incarcerated in every U.S. state prison and local jail. GAO analyzed (a) DOJ data on criminal alien incarcerations, arrests, convictions, and costs; (b) conviction data from the five state prison systems with the largest number of SCAAP criminal alien incarcerations in fiscal year 2015; and (c) DHS data on removals. For its arrest analyses, GAO selected generalizable random samples of (1) 500 criminal aliens from about 197,000 that were incarcerated in federal prisons from fiscal years 2011 through 2016 and (2) 500 SCAAP criminal aliens from about 533,000 that were incarcerated in state prisons and local jails from fiscal years 2010 through 2015. These samples included only those criminal aliens who had a Federal Bureau of Investigation (FBI) number—a unique identifier. This unique identifier allowed GAO to obtain arrest/transfer histories for these criminal aliens from FBI's database, which includes data from law enforcement agencies across the nation. While the samples selected for the arrest analyses allowed GAO to estimate and provide insights about the arrest history of the criminal aliens in the study populations, these findings are not generalizable to the arrest history of criminal aliens not included in these populations. These data did not allow GAO to distinguish between a new arrest and a transfer from one agency to another; therefore, these are collectively referred to as "arrests/transfers." An arrest/transfer can be for multiple offenses. GAO's arrest analyses have a margin of error of plus or minus 7 percentage points or fewer. For GAO's analyses of state conviction data, information obtained from the selected state prison systems is not generalizable to all state prison systems, but provides useful insights on why criminal aliens were incarcerated.

Chapter 4 - Facilities that produce, use, or store hazardous chemicals could be targeted or used by terrorists to inflict mass casualties, damage, and fear. DHS established the CFATS program to assess the risk posed by these facilities and inspect them to ensure compliance with DHS standards. DHS places high-risk facilities in risk-based tiers and is to conduct inspections after it approves their security plans. Under the CFATS Act of 2014, authorization for the CFATS program expires in January 2019. GAO assessed the extent to which DHS has (1) enhanced the process for identifying high-risk facilities and assigning them to tiers, (2) conducted facility inspections and measured facility security, and (3) ensured that information is shared with emergency responders to prepare them for incidents at high-risk facilities. GAO reviewed DHS reports and data on compliance inspections and interviewed DHS officials. GAO also obtained non-generalizable information from 11 trade associations representing chemical facilities regarding DHS outreach and from 15 emergency planning committees about their awareness of CFATS and the chemicals it covers.

Chapter 5 - Misconduct is generally considered an action by an employee that impedes the efficiency of the agency's service or mission. Misconduct incidents can affect other aspects of employee morale and performance and impede an agency's efforts to achieve its mission. GAO was asked to examine how executive branch agencies address employee misconduct. This report (1) describes the process agencies are required to follow in responding to employee misconduct; (2) identifies alternative approaches to the formal process that agencies can use and assesses what factors affect agencies' responses to misconduct; (3) describes trends in removals and other adverse actions resulting from misconduct; and (4) identifies key practices agencies can use to help them better prevent and address misconduct. To address these objectives, GAO reviewed relevant sections of title 5 of the U.S.C; analyzed MSPB and OPM data, and interviewed, among others, agency officials and subject-matter experts.

In: Top 5 Reports and Testimonies … ISBN: 978-1-53615-511-2
Editors: C. Johnston and W. Payne

Chapter 1

2020 CENSUS: CONTINUED MANAGEMENT ATTENTION NEEDED TO ADDRESS CHALLENGES AND RISKS WITH DEVELOPING, TESTING, AND SECURING IT SYSTEMS*

United States Government Accountability Office

ABBREVIATIONS

Bureau	U.S. Census Bureau
CEDCaP	Census Enterprise Data Collection and Processing
CIO	chief information officer
Commerce	Department of Commerce

* This is an edited, reformatted and augmented version of the United States Government Accountability Office Report to Congressional Requesters, Publication No. GAO-18-655, dated August 2018.

FISMA	Federal Information Security Management Act of 2002 and Federal Information Security Modernization Act of 2014
FITARA	Federal Information Technology Acquisition Reform Act
IT	information technology
NIST	National Institute of Standards and Technology
OMB	Office of Management and Budget
POA&M	plan of action and milestones
PII	personally identifiable information

WHY GAO DID THIS STUDY

One of the Bureau's most important functions is to conduct a complete and accurate decennial census of the U.S. population. The decennial census is mandated by the Constitution and provides vital data for the nation. The Bureau plans to significantly change the methods and technology it uses to count the population with the 2020 Census, such as by offering an option for households to respond to the survey via the Internet. In preparation for the 2020 Census, the Bureau is conducting a test of all key systems and operations (referred to as the 2018 End-to-End Test), which began in August 2017 and runs through April 2019.

GAO was asked to review the Bureau's IT readiness for the 2020 Census. This chapter (1) determines the Bureau's progress in developing and testing systems for the 2018 End-to-End Test and (2) describes the challenges and risks that the Bureau has faced in implementing and securing these systems. To do this, GAO reviewed key documentation, including plans for system development and testing, and outcomes of key IT milestone reviews and security assessments.

What GAO Recommends

Over the past decade, GAO has made 93 recommendations specific to the 2020 Census to address the issues raised in this and other products. As of August 2018, the Bureau had implemented 61 recommendations, and had taken initial steps—including developing action plans—to implement the other 32 recommendations. GAO is not making additional recommendations in this chapter.

What GAO Found

The Census Bureau (Bureau) has continued to make progress in developing and testing information technology (IT) systems for the 2020 Census. Specifically, as of June 2018, the Bureau had completed all development activities for 36 of the 44 systems needed to support the 2018 End-to-End Test, and was in the process of completing these activities for the remaining 8 systems. In addition, the Bureau had completed all system and integration testing activities for 20 of the 44 systems, and was in the process of conducting these activities for the remaining 24 systems.

Nevertheless, the Bureau continues to face significant challenges and risks in its efforts to manage the schedules, contracts, costs, and cybersecurity of its 2020 Census systems.

- *Schedule management*: The Bureau's schedule for developing systems to support the 2018 End-to-End Test has experienced delays. These delays have compressed the time available for system and integration testing, and several systems experienced problems during the 2018 End-to-End Test. In addition, the Bureau is currently revising the system development and testing schedule for the 2020 Census as a result of challenges experienced and lessons learned while completing these activities during the 2018 End-to-End Test. Continued schedule management challenges may compress the time available for the remaining system and integration testing and increase the risk that systems will not function as intended.

- *Contractor oversight*: Among other challenges, the Bureau is still filling vacancies in the government program management office that is overseeing its key integration contractor. In June 2018, Bureau officials reported that 33 of the office's 58 federal employee positions were vacant. This adds risk that the office may not be able to provide adequate oversight of contractor cost, schedule, and performance.
- *IT cost growth*: The Bureau reported that its estimated IT costs had grown from $3.41 billion in October 2015 to $4.97 billion in December 2017—an increase of $1.56 billion. This increase was due, in large part, to the addition of technical integration services and updated costs for other major contracts (such as the contract for mobile devices). The amount of cost growth since the October 2015 estimate raises questions as to whether the Bureau has a complete understanding of the IT costs associated with the 2020 Census.
- *Cybersecurity*: The Bureau has made progress by completing the security assessments for 33 of the 44 systems needed to support the 2018 End-toEnd Test. However, as of June 2018, the Bureau had identified nearly 3,100 security weaknesses that will need to be addressed in the coming months. Because the 2020 Census involves collecting personal information from over a hundred million households across the country, it will be important that the Bureau addresses system security weaknesses in a timely manner and ensures that risks are at an acceptable level before systems are deployed.

With the 2020 Census less than 2 years away, it is critical that the Bureau address these challenges and risks to ensure that its IT systems are developed, tested, and secured in time to support the count of the nation's population.

August 30, 2018
Congressional Requesters

Conducting the decennial census of the U.S. population is mandated by the Constitution and provides vital data for the nation. The information that the census collects is used to apportion the seats of the House of Representatives; redraw congressional districts; allocate billions of dollars each year in federal financial assistance; and provide a social, demographic, and economic profile of the nation's people to guide policy decisions at each level of government. Further, businesses use census data to market new services and products and to tailor existing ones to demographic changes.

For 2020, a complete count of the nation's population is an enormous undertaking. The U.S. Census Bureau (Bureau), a component of the Department of Commerce (Commerce), is seeking to control the cost of the census while it implements several innovations and manages the processes of acquiring and developing information technology (IT) systems. However, in recent years, we have identified challenges that raise serious concerns about the Bureau's ability to conduct a cost-effective count of the nation, including issues with the agency's research, testing, planning, scheduling, cost estimation, systems development, and cybersecurity practices. We also added the 2020 Census to GAO's high-risk list in February 2017.[1]

Currently, the Bureau is conducting the 2018 End-to-End Test, which began in August 2017 and runs through April 2019. This effort is the Bureau's final opportunity to test all key systems and operations in a census-like environment to ensure readiness for the 2020 Census.

Given the importance of the IT systems to the 2018 End-to-End Test and the 2020 Census, you asked us to review the Bureau's IT readiness for the 2020 Census. Our specific objectives were to (1) determine the

---

[1] GAO, *High-Risk Series: Progress on Many High-Risk Areas, While Substantial Efforts Needed on Others*, GAO-17-317 (Washington, D.C.: Feb. 15, 2017). GAO maintains a high-risk program to focus attention on government operations that it identifies as high risk due to their greater vulnerabilities to fraud, waste, abuse, and mismanagement or the need for transformation to address economy, efficiency, or effectiveness challenges.

progress that the Bureau has made in developing and testing the critical IT systems for the 2018 End-to-End Test, and (2) describe the challenges and risks that the Bureau has faced in implementing and securing these systems.

To determine the progress in developing and testing systems for the 2018 End-to-End Test, we updated information related to our recent work on the Bureau's system development and testing efforts.[2] To do this, we reviewed relevant Bureau documentation on the 44 IT systems in the 2018 End-to-End Test, including the 2020 Census Operational Plan, the Bureau's system integration and implementation plan, outcomes of key IT milestone reviews, and requirements traceability matrices. We also interviewed agency officials, including the Chief Information Officer (CIO), the Associate Director for Decennial Census Programs, and officials from the Bureau's technical integration contractor.

To describe the challenges and risks that the Bureau has faced in implementing and securing these systems, we reviewed Bureau documentation, including risk and issue registers, meeting minutes from 2020 Census executive review meetings, and presentations from bimonthly meetings from the Bureau's technical integration contractor. We also reviewed documentation discussing the Bureau's progress in mitigating or addressing the risks and challenges, including plans for 2020 Census system development and testing, deliverables from key contractors (including the technical integration contractor), the basis of estimate documentation for the 2020 Census cost estimate developed in December 2017, results of 2020 Census executive review board meetings, outcomes

---

[2] For example, GAO, 2020 Census: Actions Needed to Mitigate Key Risks Jeopardizing a Cost-Effective and Secure Enumeration, GAO-18-543T (Washington, D.C.: May 8, 2018); 2020 Census: Continued Management Attention Needed to Mitigate Key Risks Jeopardizing a Cost-Effective and Secure Enumeration, GAO-18-416T (Washington, D.C.: Apr. 18, 2018); 2020 Census: Actions Needed to Mitigate Key Risks Jeopardizing a Cost-Effective Enumeration, GAO-18-215T (Washington, D.C.: Oct. 31, 2017); 2020 Census: Continued Management Attention Needed to Oversee Innovations, Develop and Secure IT Systems, and Improve Cost Estimation, GAO-18-141T (Washington, D.C.: Oct. 12, 2017); 2020 Census: Sustained Attention to Innovations, IT Systems, and Cost Estimation Is Needed, GAO-17-584T (Washington, D.C.: May 3, 2017); and Information Technology: Better Management of Interdependencies between Programs Supporting 2020 Census Is Needed, GAO-16-623 (Washington, D.C.: Aug. 9, 2016).

from security assessments, and the Bureau's list of plans of action and milestones (POA&Ms). In addition, we interviewed relevant agency officials, including the CIO and the Chief Information Security Officer.

We conducted this performance audit from August 2016 to August 2018 in accordance with generally accepted government auditing standards. Those standards require that we plan and perform the audit to obtain sufficient, appropriate evidence to provide a reasonable basis for our findings and conclusions based on our audit objectives. We believe that the evidence obtained provides a reasonable basis for our findings and conclusions based on our audit objectives.

BACKGROUND

For the 2020 Census, the Bureau is significantly changing how it intends to conduct the census, in part by re-engineering key census-taking methods and infrastructure, and making use of new IT applications and systems. For example, the Bureau plans to offer an option for households to respond to the survey via the Internet and enable field-based enumerators[3] to use applications on mobile devices to collect survey data from households. In December 2017, the Bureau estimated the total cost of the redesigned 2020 Census to be about $15.6 billion, more than $3 billion higher than it estimated in October 2015.

To inform the design of the 2020 Census operations and systems, the Bureau held several major operational tests. These tests included (1) the 2016 Census tests in Texas and California, which evaluated, among other things, the efficiency of non-response follow-up[4] using contractor-provided mobile devices; and (2) the 2017 Census Test—a nationwide sample of how individuals respond to Census questions using paper, the Internet, or

[3] Enumerators are Census Bureau employees who travel from door-to-door throughout the country to try to obtain census data from individuals who do not respond through other means, including the Internet, on paper, or by phone.

[4] In non-response follow-up, if a household does not respond to the census by a certain date, the Bureau will conduct an in-person visit by an enumerator to collect census data using a mobile device provided by the Bureau.

the phone—which evaluated key new IT components, such as the Internet self-response system and the use of a cloud-based infrastructure.

Currently, the Bureau is conducting the 2018 End-to-End Test, which began in August locations—Rhode Island, West Virginia, and Washington state. The Bureau is currently testing its non-response operation in Rhode Island.[5] This operation began in May 2018 and is scheduled to conclude in August 2018.

Bureau Plans to Rely Heavily on IT for the 2018 End-to-End Test and the 2020 Census

The Bureau plans to rely heavily on both new and legacy IT systems and infrastructure to support the 2018 End-to-End Test and the 2020 Census operations. For example, the Bureau's plans call for deploying and using 44 systems in the 2018 End-to-End Test. Eleven of these systems are currently being developed or modified as part of an enterprise-wide initiative called Census Enterprise Data Collection and Processing (CEDCaP).[6] This initiative is a large and complex modernization program intended to deliver a system-of-systems to support all of the Bureau's survey data collection and processing functions, rather than continuing to rely on unique, survey-specific systems with redundant capabilities.

As part of the 2018 End-to-End Test, the Bureau plans to incrementally test, deploy, and use the 44 systems from December 2016 through the end of the test in April 2019.[7] These systems are to be deployed in 14 groups based on the operations that they support in the test,

---

[5] The Bureau originally planned to perform the entire 2018 End-to-End Test in all three locations. However, in May 2017, the Bureau scaled back the operational scope of the test and, of the three planned test sites, the Bureau would fully implement the test in the Rhode Island site only.

[6] The Bureau is pursuing enterprise-wide technology solutions intended to support other major surveys the Bureau conducts as well, such as the American Community Survey and the Economic Census.

[7] Prior to deployment in the 2018 End-to-End Test, the systems are to undergo various forms of testing, including system-level testing to ensure that the system meets business requirements, and integration testing to validate (among other things) the interfaces between systems.

including address canvassing, self-response (i.e., Internet, phone, or paper), field enumeration, and tabulation and dissemination. A system may be deployed for one operation in the 2018 End-to-End Test (such as address canvassing), and then be deployed again for a subsequent operation in the test (such as field enumeration).

Following the 2018 End-to-End Test, the Bureau has additional system development and testing activities planned leading up to the 2020 Census. Specifically, Bureau officials said they expect that the 44 systems used in the 2018 End-to-End Test will need to undergo further development and testing due to, among other things, the need to add functionality that was not part of the end-to-end test, scale system performance to support the number of respondents expected during the 2020 Census, and to address system defects identified during the 2018 End-to-End Test. Bureau officials also reported that they plan to use an additional 8 systems that were not included in the 2018 End-to-End Test—for a total of 52 systems—to carry out the 2020 Census operations.[8]

As of June 2018, the Bureau planned to deploy the 52 systems for the 2020 Census in four groups, or operational releases, to support the key operations in the 2020 Census: (1) recruiting and hiring; (2) address canvassing; (3) self-response, non-response follow-up, and fraud detection; and (4) reporting and coverage measurement.[9] The systems are grouped according to the operations that they support during the 2020 Census. For example, the third operational release—which includes the most systems—has 47 systems to be used for self-response (including via the Internet), non-response follow-up, and fraud detection.

In addition to its systems and applications, the Bureau is designing, configuring, and managing the IT infrastructure needed for the 2020 Census. The Bureau's 2020 Census infrastructure includes: (1) an on premises data center being managed by the Bureau; (2) an on premises data

[8] Several of these systems are for the coverage measurement operation, which was removed from the scope of the 2018 End-to-End Test. Coverage measurement evaluates the quality of the census data by estimating the census coverage based on a post-enumeration survey.

[9] Similar to the 2018 End-to-End Test, a system being used in the 2020 Census may be deployed multiple times (with additional or new functionality) if that system is needed for more than one of these operations.

center being managed by the Bureau's technical integration contractor; (3) a cloud-based infrastructure using Amazon Web Services being managed by the technical integration contractor; (4) hardware located at the National Processing Center for printing and mailing systems; and (5) hardware managed and located elsewhere for those systems that are provided as software-as-a-service.[10]

THE BUREAU HAS MADE PROGRESS IN DEVELOPING AND TESTING ITS IT SYSTEMS FOR THE 2018 END-TO-END TEST

The Bureau has made progress in developing and testing the IT systems for the 2018 End-to-End Test. In this regard, it has finished developing a majority of the 44 systems in the test. Specifically, as of June 2018, the Bureau had completed all development activities for 36 of the 44 systems in the test, and was in the process of completing these activities for the remaining 8 systems. Figure 1 summarizes the development status for the 44 systems planned for the 2018 End-to-End Test, as of June 2018.

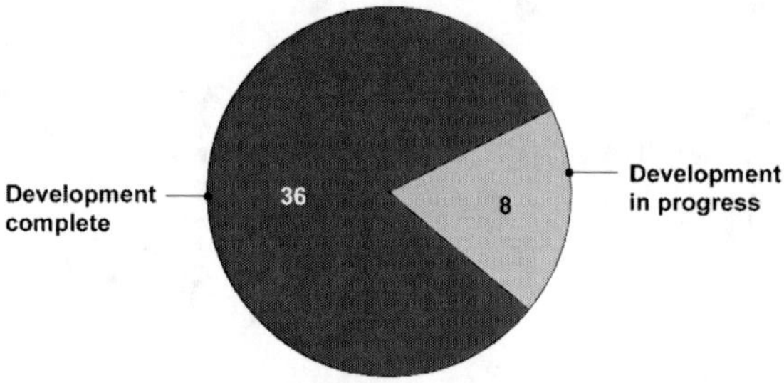

Source: GAO analysis of Census Bureau data. | GAO-18-655.

Figure 1. Development Status for the 44 Systems in the Census Bureau's 2018 End-to-End Test, as of June 2018.

---

[10] Software-as-a-service involves purchasing commercial software that is operated and maintained by a commercial vendor. Examples of systems that the Bureau has acquired as software-as-a-service include Census Questionnaire Assistance (the phone response system), and Recruiting and Assessment.

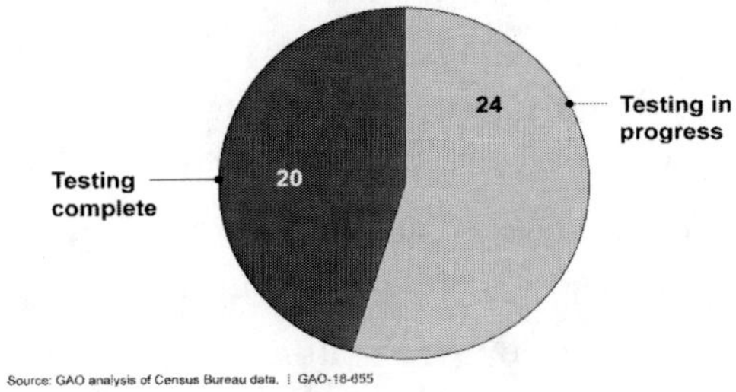

Source: GAO analysis of Census Bureau data. | GAO-18-655.

Figure 2. Testing Status for the 44 Systems in the Census Bureau's 2018 End-to-End Test, as of June 2018.

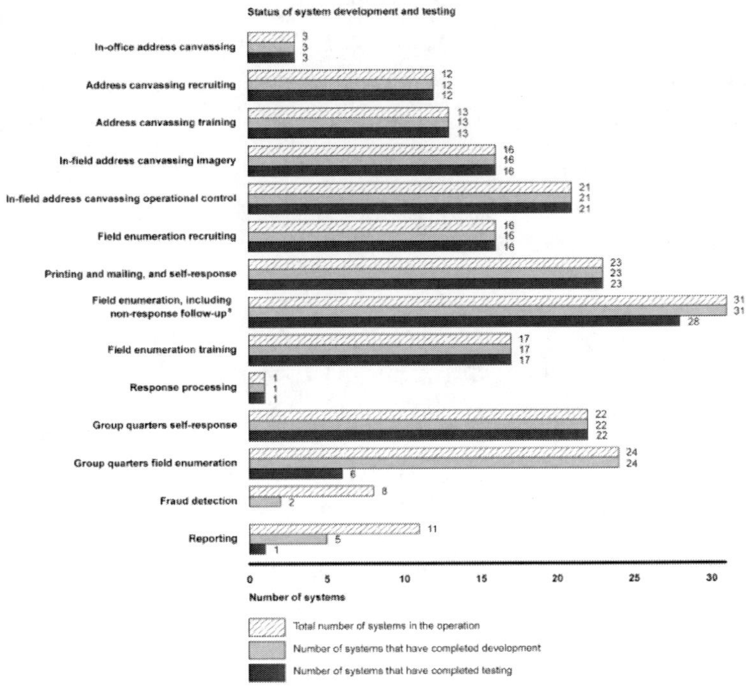

Source: GAO analysis of Census Bureau data. | GAO-18-655.

[a] In addition to non-response follow-up, field enumeration also includes operations such as update/leave and coverage improvement. In update/leave, listers update a housing unit's address and leave a questionnaire to allow the household to self-respond. The goal of coverage improvement is to resolve erroneous enumerations (such as people counted in the wrong place or more than once) and omissions.

Figure 3. Development and Testing Status for System Operations in the Census Bureau's 2018 End-to-End Test, as of June 2018.

In addition, as of June 2018, the Bureau had completed all testing activities (e.g., system and integration testing) for 20 of the 44 systems included the 2018 End-to-End Test, and was in the process of conducting these tests for the remaining 24 systems. Figure 2 summarizes the status of testing for the 44 systems in the 2018 End-to-End Test. Further, appendix I provides additional details about the status of the development and testing activities for these systems.

In total, as of June 2018, the Bureau had completed all development and testing activities to support 10 of the 14 operations in the 2018 End-to-End Test, such as in-field address canvassing and response processing.[11] For the remaining 4 operations—field enumeration, group quarters enumeration, fraud detection, and reporting—system development and/or testing activities were in process, but had not been completed. Figure 3 depicts the total number of systems supporting each operation, the number of systems that have completed development and testing, and the status of the operation.

THE BUREAU CONTINUES TO FACE CHALLENGES AND RISKS IN IMPLEMENTING ITS IT SYSTEMS FOR THE 2020 CENSUS

Even as the Bureau has made progress in its system development and testing activities, it continues to face challenges in managing and overseeing the development and testing of its IT systems for the 2018 End-to-End Test and the 2020 Census. Specifically, we have noted challenges in the Bureau's efforts to manage the schedules, contracts, costs, governance and internal coordination, and cybersecurity of its systems.

---

[11] As stated previously, the 44 systems in the test are to be deployed multiple times in a series of operations (such as group quarters enumeration). That is, a system may be deployed for one operation in the 2018 End-to-End Test (such as address canvassing), and be deployed again for a subsequent operation in the test (such as group quarters enumeration). As such, additional development and testing is to occur each time a system is deployed.

Schedule Management

The Bureau has faced significant challenges in managing its schedule for developing and testing systems for the 2018 End-to-End Test. Further, due, in part, to these challenges, the Bureau is replanning key IT milestones for the 2020 Census.

In May 2018, we reported that the Bureau had delayed by several months key IT milestone dates (e.g., dates to begin system integration testing) for a majority of the 14 operations in the 2018 End-to-End Test.[12] For example, the Bureau moved the test readiness review date for the fraud detection operation from April 2018 to July 2018—a delay of 3 months. These delays have compressed the time the Bureau has had for integration testing before the systems are deployed in the 2018 End-to-End Test.

Several of the systems subsequently experienced problems during the end-to-end test, including the mobile device applications being used by enumerators for the non-response follow-up[13] operation of the test.[14] For example, Bureau officials reported that enumerators have experienced problems with the sensitivity of the mobile devices' touch screen. More specifically, in certain cases, the mobile device application did not identify that the enumerator had made a selection on the touch screen until after the enumerator attempted to select it multiple times.

In addition, we previously reported that the delays in system development and testing had reduced the time available to conduct the security reviews and approvals for the systems being used in the 2018 End-to-End Test.[15] Officials in the Bureau's Office of Information Security stated that the original plan was to have at least 6 to 8 weeks to perform security assessments for each system. However, given the compressed time frames,

---

[12] GAO-18-543T.

[13] In non-response follow-up, if a household does not respond to the census by a certain date, the Bureau will send out employees to visit the home. The Bureau's plan is for these enumerators to use a census application, on a mobile device provided by the Bureau, to capture the information given to them by the in-person interviews.

[14] As mentioned previously, the non-response follow-up operation of the End-to-End Test is being performed in Rhode Island and began in May 2018 and is scheduled to conclude in August 2018.

[15] GAO-18-543T.

Bureau officials informed us that, in some instances, they have had 5 to 8 days to complete certain assessments. This resulted in systems being deployed before the security of all system components were assessed.[16] We concluded that, going forward, it would be important for these security assessments to be completed in a timely manner and that risks be at an acceptable level before the systems are deployed.

Due in part to IT development and testing schedule challenges that it has identified during the 2018 End-to-End Test, the Bureau is in the process of revising the milestone dates for the additional system development and testing that is to occur after the 2018 End-to-End Test and before the 2020 Census. As noted earlier, the Bureau plans to develop, test, and deploy the 52 systems in the 2020 Census in four operational releases.

According to the Bureau's plans, the agency originally planned to complete development for its first 2020 Census operational release (for recruiting and hiring) in May 2018. However, in June 2018, Bureau officials reported that the Bureau did not meet the May 2018 delivery date for the 2020 Census recruiting and hiring operational release. Additionally, the agency originally planned to complete integration and testing for this operational release by July 2018. However, in July 2018, Bureau officials reported that this milestone had been delayed to August 2018. The Bureau's original milestone dates for the operational releases, reflecting the system development completion status for recruiting and hiring, are shown in table 1.

Bureau officials reported that they intend to revise the development and testing milestone dates for all four operational releases for the 2020 Census, but that they do not expect the final deployment dates to change. The officials further noted that they are planning to incorporate lessons learned to date from the 2018 End-to-End Test as part of the replanning efforts. However, Bureau officials had not yet identified a specific time frame for completing these efforts.

---

[16] According to the Bureau's Chief Information Security Officer, components that do not have all controls assessed are to be tracked until the assessments are completed, even if it is after the system deploys.

Table 1. The Census Bureau's Original Milestone Dates for Operational Releases for the 2020 Census, as of July 2018

Operational release name	Number of systems in the operational release	Expected completion date for system development	Expected completion date for integration and test	Expected deployment date
1. Recruiting and hiring	21	May 2018 (Not met)	July 2018 (Not met)	September 2018
2. Address canvassing	29	November 2018	March 2019	May 2019
3.Self-response, non-response follow-up, and fraud detection	47	February 2019	June 2019	November 2019
4.Reporting and coverage measurement	25	October 2019	February 2020	July 2020

Source: GAO analysis of Census Bureau data. | GAO-18-655.

Managing the schedule for system development and testing is also important because the Bureau plans to conduct system performance and scalability testing after the 2018 End-to-End Test and prior to the 2020 Census.[17] Specifically, in February 2018, the Bureau established an approach to conducting performance and scalability testing that began in February 2018 (with system design reviews and analyses) and is expected to be completed in October 2019.

As of July 2018, the Bureau reported that it had completed design reviews and analysis for 11 systems (such as the operational control system) and had developed performance test plans for 8 systems. Figure 4 summarizes the Bureau's performance and scalability test plans.

As planning for the 2020 Census continues, it will be important for the Bureau to provide adequate time for system development and testing activities. This will help ensure that the time available for security assessments is not reduced as it has been, thus far, during the 2018 End-

[17] Performance testing is the process of determining how a system behaves under a specific workload (e.g., number of users). Scalability testing is a subset of performance testing to determine a system's effectiveness in handling an increasing number of users.

To-End Test. We have previously reported that, without adequate time for completing these security assessments, the Bureau will be challenged in ensuring that risks are at an acceptable level before the systems are deployed for the 2020 Census.

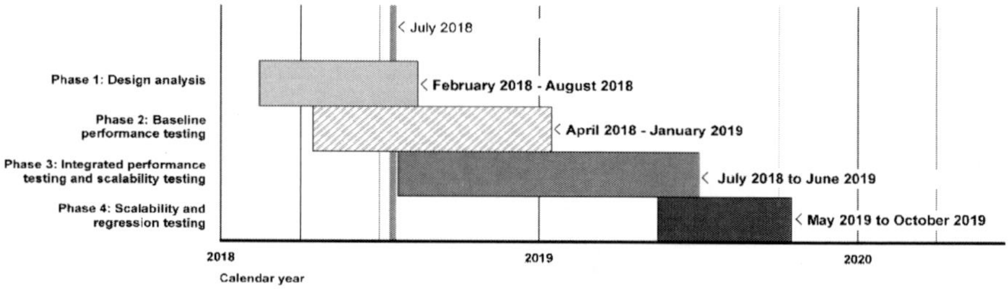

Source: GAO analysis of Census Bureau data. | GAO-18-655.

Figure 4. The Census's Bureau's Expected Performance and Scalability Test Plans for the 2020 Census, as of July 2018.

Contract Management

The Bureau also faces challenges in managing its significant contractor support. The Bureau is relying on contractor support in many areas to prepare for the 2020 Census. For example, it is relying on contractors to develop a number of systems and components of the IT infrastructure. These activities include (1) developing the IT platform (as part of the CEDCaP program) that is intended to be used to collect data from households responding via the Internet and telephone, and for non-response follow-up activities; (2) procuring the mobile devices and cellular service to be used for non-response follow-up; and (3) deploying the IT and telecommunications hardware in the field offices. According to Bureau officials, contractors are also providing support in areas such as fraud detection, cloud computing services, and disaster recovery.

In addition to the development of technology, the Bureau is relying on a technical integration contractor to integrate all of the key systems and infrastructure. The Bureau awarded a contract to integrate the 2020 Census systems and infrastructure in August 2016. The contractor's work was to include evaluating the systems and infrastructure and acquiring the infrastructure (e.g., cloud or data center) to meet the Bureau's scalability and performance needs. It was also to include integrating all of the systems, supporting technical testing activities, and developing plans for ensuring the continuity of operations. Since the contract was awarded, the Bureau has modified the scope to also include assisting with operational testing activities, conducting performance testing for two Internet self-response systems, and providing technical support for the implementation of the paper data capture system.

However, the Bureau continues to face staffing challenges that could impact its ability to manage and oversee the technical integration contractor. Specifically, the Bureau is managing the integration contractor through a government program management office, but this office is still filling vacancies. In June 2018, Bureau officials reported that 33 of the office's 58 federal employee positions were vacant. This means that the Bureau has only filled 2 of these positions since we originally reported about this risk in October 2017. These vacancies increase the risk that the program management office may not be sufficiently staffed to provide adequate oversight of contractor cost, schedule, and performance.

The development and testing schedule delays during the preparations for the 2018 End-to-End Test raise concerns about the Bureau's ability to effectively perform contractor management. As we reported in November 2016, a greater reliance on contractors for these components of the 2020 Census requires the Bureau to focus on sound management and oversight of the key contracts, projects, and systems.[18]

It Cost Growth

The Bureau faces challenges in controlling IT cost growth. Specifically, the Bureau's October 2015 cost estimate included about $3.41 billion in total IT costs for fiscal years 2012 through 2023. These included

[18] GAO, Information Technology: Uncertainty Remains about the Bureau's Readiness for a Key Decennial Census Test, GAO-17-221T (Washington, D.C.: Nov. 16, 2016).

costs for, among other things, system engineering, test and evaluation, and infrastructure, as well as for a portion of the CEDCaP program.[19] However, in October 2017, we reported[20] that IT costs would likely be at least $4.8 billion, due to increases in costs associated with the CEDCaP program[21] and certain IT contracts (including those associated with technical integration and mobile devices).

In December 2017, the Bureau reported that its estimated IT costs had grown from $3.41 billion to $4.97 billion—an increase of $1.56 billion. Figure 5 identifies the Bureau's estimate of total IT costs associated with the 2020 program as of December 2017.

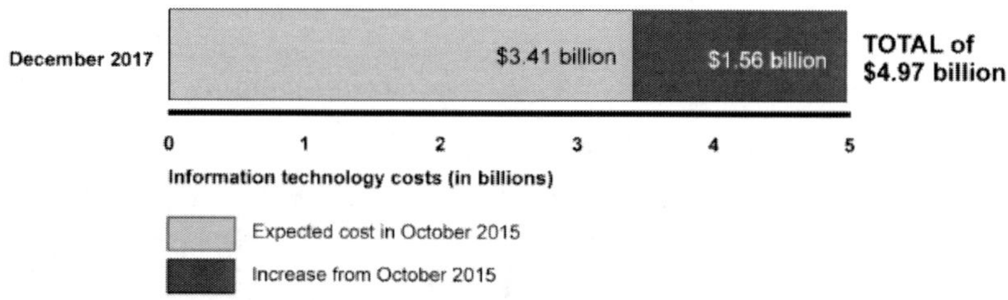

Source: GAO analysis of Census Bureau data. | GAO-18-655.

Figure 5. Total Information Technology Costs Estimated by the Census Bureau, as of December 2017.

The cost increases were due, in large part, to the Bureau (1) updating the cost estimate for the CEDCaP program, (2) including an estimate for technical integration services, and (3) updating costs related to other major contracts (such as mobile device as a service).[22] Table 2 describes the IT costs that comprised the Bureau's cost estimate as of December 2017.

---

[19] The 2020 Census program pays for the portion of costs for the CEDCaP program that relate to 2020 Census operations. According to the October 2015 estimate, the portion of CEDCaP costs associated with the 2020 Census was estimated at $328 million of the $548 million total program estimate.

[20] GAO-18-215T.

[21] In May 2017, the Bureau reported that the CEDCaP program's cost estimate was increasing by about $400 million—from its original estimate of $548 million in 2013 to a revised estimate of $965 million in May 2017.

[22] As part of mobile device as a service, the Bureau plans to provide mobile devices (including mobile phones, tablets, and laptops) and cellular service to field staff to support operations such as address canvassing and non-response follow-up.

Table 2. Total 2020 Census Information Technology (IT) Costs Estimated by the Census Bureau, by Cost Category, as of December 2017

IT cost category	Expected cost
Technical integration services	$1,492
Census questionnaire assistance	$817
Other IT services, such as day-to-day IT support	$808
Census Enterprise Data Collection and Processing (CEDCaP) costs related to the 2020 Census	$509[a]
Decennial device as a service	$489
Field IT deployment	$450
Other non-CEDCaP systems, such as recruitment and personnel systems	$401
Total	$4,966

Source: GAO analysis of Census Bureau data. | GAO-18-655.

[a] The 2020 program pays for a portion of the costs for the CEDCaP program. As of May 2017, the Census Bureau estimated that the entire cost of the CEDCaP program would be about $965 million.

IT cost information that is accurately reported and clearly communicated is necessary to help ensure that Congress and the public have confidence that taxpayer funds are being spent in an appropriate manner. However, the amount of cost growth since the October 2015 estimate raises questions as to whether the Bureau has a complete understanding of the IT costs associated with the 2020 program. We have ongoing work reviewing the extent to which the Bureau's December 2017 cost estimate is reliable.

Governance and Internal Coordination

Effective governance can drive change, provide oversight, and ensure accountability for results. Further, effective IT governance was envisioned in the statutory provisions enacted in 2014 and referred to as the Federal Information Technology Acquisition Reform Act (FITARA),[23] which strengthened and reinforced the role of the departmental CIO. The

[23] Carl Levin and Howard P. 'Buck' McKeon National Defense Authorization Act for Fiscal Year 2015, Pub. L. No. 113-291, div. A, title VIII, subtitle D, 128 Stat. 3292, 3438-50 (Dec. 19, 2014).

component CIO (such as the Bureau's CIO) also plays a role in effective IT governance, as the component is subject to the oversight and policies of the parent department implementing FITARA.

Our work has noted that officials in Commerce's Office of the Secretary have increased their oversight of the Bureau's preparations for the 2020 Census by holding regular meetings to discuss contracts, expected costs, and risks, among other topics.[24] For example, Bureau officials in the Decennial Directorate told us that they have recently begun meeting with the Secretary of Commerce on a monthly basis, and with the Under Secretary of Commerce for Economic Affairs on a weekly basis, to discuss 2020 Census issues. Moreover, the department's Acting CIO has also been involved in overseeing the Bureau's IT system readiness.

The Bureau has also appointed two new assistant directors within the Decennial Directorate. Each of these individuals is responsible for overseeing aspects of the 2020 Census program, to include schedules, contracts, and system development. In addition, Bureau officials told us that the CIO (or a designated representative) is to be a member of the governance boards that oversee all of the operations and technology for the 2020 Census, in order to ensure executive-level oversight of the key systems and technology.

Nevertheless, in August 2016, we reported on challenges that the Bureau has had with IT governance and internal coordination, including weaknesses in its ability to monitor and control IT project costs, schedules, and performance.[25] We made eight recommendations to the Secretary of Commerce to direct the Bureau to, among other things, better ensure that risks are adequately identified and schedules are aligned. The department agreed with our recommendations. As of June 2018, the Bureau had fully implemented five of the recommendations and had taken initial steps toward implementing the other three recommendations.

Given the schedule delays and cost increases previously mentioned, and the vast amount of development, testing, and security assessments left to be completed, we remain concerned about executive-level oversight of systems

[24] GAO-18-543T.
[25] GAO-16-623.

and security. Moving forward, it will be important that the CIO and other Bureau executives continue to use a collaborative governance approach to effectively manage risks and ensure that the IT solutions meet the needs of the agency within cost and schedule.

Cybersecurity

In November 2016, we described the significant challenges that the Bureau faced in securing systems and data for the 2020 Census, and we noted that tight time frames could exacerbate these challenges.[26] Two such challenges were (1) ensuring that individuals gain only limited and appropriate access to the 2020 Census data, including personally identifiable information (PII), such as name, personal address, and date of birth; and (2) making certain that security assessments were completed in a timely manner and that risks were at an acceptable level.[27] Protecting PII, for example, is especially important because a majority of the 44 systems to be used in the 2018 End-to-End Test contain such information, as reflected in figure 6.[28]

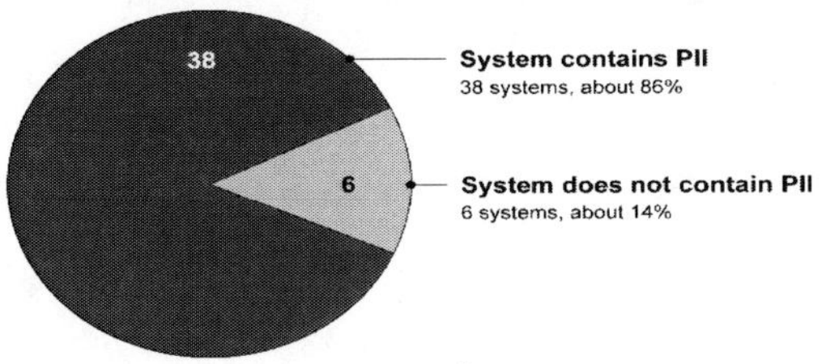

Source: GAO analysis of Census Bureau data. | GAO-18-655.

Figure 6. Personally Identifiable Information (PII) in Census Bureau Systems Included in the 2018 End-to-End Test, as of June 2018.

[26] GAO-17-584T.
[27] GAO-17-221T.
[28] According to officials in the Bureau's Office of Information Security, 26 systems contain data that is protected from disclosure under Title 13 of the U.S. Code. This law protects information provided by the public for the Bureau's censuses and surveys and requires that the Bureau keep it confidential. 13 U.S.C. § 9. The Bureau may not disclose or publish any private information that identifies an individual or business, such as names, addresses, Social Security numbers, and telephone numbers.

To address these and other challenges, federal law specifies requirements for protecting federal information and information systems, such as those systems to be used in the 2020 Census. Specifically, the Federal Information Security Management Act of 2002 and the Federal Information Security Modernization Act of 2014 (FISMA) require executive branch agencies to develop, document, and implement an agency-wide program to provide security for the information and information systems that support operations and assets of the agency.[29]

Accordingly, the National Institute of Standards and Technology (NIST) developed risk management framework guidance for agencies to follow in developing information security programs.[30] In addition, the Office of Management and Budget's (OMB) revised Circular A-130 on managing federal information resources required agencies to implement the NIST risk management framework to integrate information security and risk management activities into the system development life cycle.[31]

In accordance with FISMA, NIST guidance, and OMB guidance, the Bureau's Office of the CIO established a risk management framework. This framework requires system developers to ensure that each of the Bureau's systems undergoes a full security assessment, and that system developers remediate critical deficiencies. In addition, according to the framework, system developers are to ensure that each component of a system has its own system security plan that documents how the Bureau intends to implement security controls. As a result of this requirement, system developers for a single system might develop multiple system security plans which all have to be approved as part of the system's complete security documentation.

---

[29] The Federal Information Security Modernization Act of 2014, Pub. L. No. 113-283, 128 Stat. 3073 (Dec. 18, 2014) largely superseded the Federal Information Security Management Act of 2002, enacted as Title III, E-Government Act of 2002, Pub. L. No. 107-347, 116 Stat. 2899, 2946 (Dec. 17, 2002).

[30] NIST, Guide for Applying the Risk Management Framework to Federal Information Systems: A Security Life Cycle Approach, SP 800-37, Revision 1 (Gaithersburg, Md.: February 2010).

[31] OMB, Revision of OMB Circular A-130, Managing Federal Information as a Strategic Resource (Washington, D.C.: July 28, 2016).

According to the Bureau's framework, each of the 44 systems in the 2018 End-to-End Test will need to have complete security documentation (such as system security plans) and an approved authorization to operate prior to its use in the 2018 End-to-End Test.[32] In May 2018, Bureau officials reported that they had recently updated their policies related to obtaining an authorization to operate. Specifically, once a system undergoes a security assessment and receives an authorization, the system moves into continuous monitoring.[33] According to the Bureau's Chief Information Security Officer, authorized systems do not need a formal reauthorization unless the risk posture of the system changes; this could occur, for example, if the system undergoes significant new development.

As of June 2018, most of the systems in the 2018 End-to-End Test had received an authority to operate. Specifically, according to the Bureau:

- Thirty-three of the 44 systems in the test had obtained an authorization to operate and were under continuous monitoring.
- Eight systems had obtained an authorization, but will need to be reauthorized due to additional significant planned development or changes to the infrastructure environment (e.g., from a data center to a cloud-based environment).
- Three systems do not yet have an authorization to operate for the 2018 End-to-End Test.

Figure 7 summarizes the authorization to operate status for the systems being used in the 2018 End-to-End Test, as reported by the Bureau.

---

[32] According to the Bureau's framework, systems are to obtain security authorization approval from the authorizing official in order to operate. Specifically, the authorizing official evaluates the security authorization package and provides system authorization if the overall risk level is acceptable. In addition, according to the Bureau's IT security program policy, the issuance of an authorization to operate for a system requires support of both the technical authorizing official (i.e., the CIO) and the business authorizing official responsible for funding and managing the system (i.e., the Associate Director for Decennial Census Programs).

[33] According to the Bureau's risk management framework, once a system obtains an authorization, it is transitioned to the continuous monitoring process where the authorizing official can provide ongoing authorization for system operation as long as the risk level remains acceptable.

According to the Bureau's framework, security assessment findings that need to be remediated are tracked in a plan of action and milestones (POA&M). Specifically, the POA&M provides a description of the vulnerability identified in the security assessment as a result of a control weakness. As of June 2018, the Bureau had nearly 3,100 open POA&Ms to remediate issues identified during security assessments performed for the 2018 End-to-End Test. Of these nearly 3,100 POA&Ms, 43 were considered "very high risk" or "high risk" weaknesses. Further, over 2,700 of the POA&Ms were related to the infrastructure components being developed by the technical integration contractor. Officials from the Bureau and the technical integration contractor reported that they are currently working to address these POA&Ms.

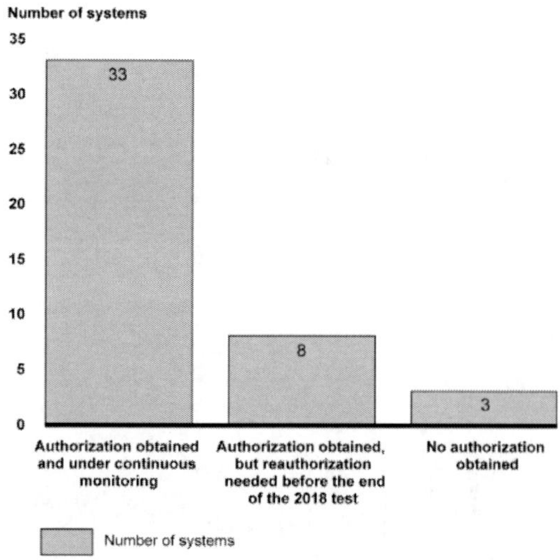

Source: GAO analysis of Census Bureau data. | GAO-18-655.

Figure 7. Authorization to Operate Status for the 44 Systems Being Used in the Census Bureau's 2018 End-to-End Test, as of June 2018.

Further, because several of the systems that will be a part of the 2018 End-to-End Test and the 2020 Census are not yet fully developed, the Bureau has not finalized all of the security controls to be implemented; assessed those controls; developed plans to remediate control weaknesses; and determined whether there is time to fully remediate any deficiencies before the systems are needed for the test. Also, as discussed earlier, the Bureau is facing system development and testing challenges that are delaying the completion of milestones and compressing the time available for security testing activities.

In addition, while the large-scale technological changes (such as Internet self-response) increase the likelihood of efficiency and effectiveness gains, they also introduce many cybersecurity challenges. The 2020 Census also involves collecting PII on over a hundred million households across the country, which further increases the need to properly secure these systems. Thus, it will be important that the Bureau provides adequate time to perform these security assessments, completes them in a timely manner, and ensures that risks are at an acceptable level before the systems are deployed.

Moving forward, the Bureau's continued attention to addressing the challenges and risks that we have identified in its efforts to manage the schedules, contracts, costs, governance and internal coordination, and cybersecurity of its IT is critical. Over the past decade, we have made 93 recommendations specific to the 2020 Census to help address, among other things, the Bureau's implementation and management of IT, scheduling, and cost estimation. The Bureau has generally agreed with those recommendations and has taken action to address 61 of them. However, as of July 2018, the remaining 32 recommendations had not been fully implemented, although the Bureau had taken initial steps— including developing action plans—to address them. Continued management attention to fully implementing these recommendations is essential for a cost-effective and secure enumeration.

AGENCY COMMENTS AND OUR EVALUATION

Commerce provided written comments on a draft of this chapter. In its comments, reprinted in appendix II, Commerce stated that it did not have any substantive disagreements with the findings in the report. The department further stated that, while it had made progress over the past year as a result of increased management focus in the areas of cost estimation, scheduling, IT security, contract management, and governance, much work remains to prepare and implement the IT systems for the 2020 Census. Commerce also provided technical comments, which we addressed, as appropriate.

Nick Marinos
Director,
Cybersecurity and Data Protection Issues

List of Requesters

The Honorable Ron Johnson
Chairman

The Honorable Claire McCaskill
Ranking Member
Committee on Homeland Security and Governmental Affairs
United States Senate

The Honorable Gary Peters
Ranking Member
Subcommittee on Federal Spending Oversight
and Emergency Management
Committee on Homeland Security and Governmental Affairs
United States Senate

The Honorable Trey Gowdy
Chairman

The Honorable Elijah E. Cummings
Ranking Member
Committee on Oversight and Government Reform
House of Representatives

Table 3. Development, Testing, and Deployment Status for the 44 Systems in the Census Bureau's 2018 End-to-End Test, as of June 2018

System name and description	Status of development	Status of testing	Expected/ actual first deployment date[a] (delay since August 2017)	Expected/ actual final deployment date[a] (delay since August 2017)	Has at least a portion of the system's functionality been deployed?
1.Block Assessment, Review and Classification Application Interactive review tool that is designed to assist an analyst in assessing a set of geographic work units.	Complete	Complete	December 2016	n/a	Yes
2. One Form Designer Plus Tool that creates paper forms including decennial questionnaires, letters, envelopes, notices of visit, language guides, and other Decennial field and public materials.	Complete	Complete	March 2017	n/a	Yes
3. Census Document System Web-based system for requesting forms design services, publications and graphics services, and printing services.	Complete	Complete	March 2017	September 2017	Yes
4. MOJO Recruiting Dashboard System that provides a dashboard to show recruiting metrics.	Complete	Complete	March 2017	June 2018	Yes

Table 3. (Continued)

System name and description	Status of development	Status of testing	Expected/ actual first deployment date[a] (delay since August 2017)	Expected/ actual final deployment date[a] (delay since August 2017)	Has at least a portion of the system's functionality been deployed?
5. Listing and Mapping Application Single instrument that enables field users to capture and provide accurate listing and mapping updates to the Master Address File/Topologically Integrated Geographic Encoding and Referencing Database.	Complete	Complete	July 2017	April 2018 (1-month delay)	Yes
6. Mobile Case Management Tool that provides mobile device-level survey case management and dashboards, and manages data transmissions and other applications on the mobile device.	Complete	Complete	July 2017	April 2018 (1-month delay)	Yes
7. MOJO Optimizer/Modeling Service to optimize the field workers' routes.	Complete	Complete	August 2017	April 2018 (1-month delay)	Yes
8. Integrated Logistics Management System System to manage logistics and resource planning.	Complete	Complete	August 2017	July 2018	Yes
9. Matching and Coding Software System that allows for clerical matching and geocoding during Non-ID Processing.	Complete	Complete	February 2018	n/a	Yes

System name and description	Status of development	Status of testing	Expected/ actual first deployment datea (delay since August 2017)	Expected/ actual final deployment datea (delay since August 2017)	Has at least a portion of the system's functionality been deployed?
10. Real Time Non-ID Processing System that matches addresses in real-time, geocodes addresses in real-time, and geo-locates housing units using web map services.	Complete	Complete	February 2018	n/a	Yes
11. Concurrent Analysis and Estimation System that stores data and uses it to execute statistical models in support of survey flow processing, analysis, and control.	Complete	Complete	March 2018	n/a	Yes
12. Enterprise Censuses and Surveys Enabling (ECaSE) – Internet Self-Response (ISR) Tool that supports self-response data collection via the Internet for respondents and by call center agents on behalf of respondents.	Complete	Complete	March 2018 (1-month delay)	n/a	Yes
13. Census Questionnaire Assistance Provides call center capability for self-response and assists respondents with responding to and completing census questionnaires.	Complete	Complete	March 2018 (1-month delay)	April 2018 (1-month delay)	Yes

Table 3. (Continued)

System name and description	Status of development	Status of testing	Expected/ actual first deployment date[a] (delay since August 2017)	Expected/ actual final deployment date[a] (delay since August 2017)	Has at least a portion of the system's functionality been deployed?
14. Census Hiring and Employment Check Administrative system that automates the clearance processing of all personnel at Census Bureau Headquarters, the Bureau of Economic Analysis, the Regional Offices, the National Processing Center, and two Computer-Assisted Telephone Interview sites.	Complete	Complete	March 2017	June 2018 (4-month delay)[b]	Yes
15. Decennial Physical Access System that is used to generate badges for certain employees, including enumerators, listers, and Census Field Supervisors.	Complete	Complete	March 2018	June 2018	Yes
16. Census Human Resources Information System Web-based personal information tool providing personnel and payroll information on desktops.	Complete	Complete	March 2017	July 2018	Yes
17. Commerce Business System that collects and reports labor hours and costs for the activities that the National Processing Center performs.	Complete	Complete	March 2017	July 2018	Yes

System name and description	Status of development	Status of testing	Expected/ actual first deployment date[a] (delay since August 2017)	Expected/ actual final deployment date[a] (delay since August 2017)	Has at least a portion of the system's functionality been deployed?
18. Decennial Service Center Suite of systems to handle all IT service requests initiated by field staff.	Complete	Complete	March 2017	July 2018	Yes
19. Desktop Services Suite of systems that includes chat.	Complete	Complete	March 2017	July 2018	Yes
20. 2020 Website for the 2018 End-to-End Test.	Complete	Complete	June 2018	January 2019	Yes
21. Unified Tracking System Data warehouse that is to combine data from a variety of Census systems, bringing the data to one place where the users can run or create reports to analyze survey and resource performance.	Complete	In progress	December 2016	July 2018	Yes
22. Master Address File/Topologically Integrated Geographic Encoding and Referencing System Database that contains, manages, and controls a repository of spatial and non-spatial data used to provide extracts to define census operations, provide maps, and support Web applications.	Complete	In progress	December 2016	October 2018	Yes

Table 3. (Continued)

System name and description	Status of development	Status of testing	Expected/ actual first deployment date[a] (delay since August 2017)	Expected/ actual final deployment date[a] (delay since August 2017)	Has at least a portion of the system's functionality been deployed?
23. Automated Tracking and Control Tool that provides customer, employee, and workflow management by automating business and support activities. It provides outbound call tracking for Geographic Partnership Programs and material tracking and check-in.	Complete	In progress	February 2018	July 2018	Yes
24. Intelligent Postal Tracking Service Mail tracking system developed by the Census Bureau and the U.S. Postal Service to trace individual mail pieces during transit.	Complete	In progress	February 2018	July 2018	Yes
25. Decennial Applicant, Personnel and Payroll Systems System that supports personnel and payroll administration for temporary, intermittent Census Bureau employees participating in the 2018 End-to-End Test.	Complete	In progress	March 2017	July 2018 (11-month delay)[b]	Yes
26. Recruiting and Assessment Tool that provides capabilities for applicant recruiting and the applicant pre-selection assessment process.	Complete	In progress	March 2017	July 2018 (5-month delay)[b]	Yes
27. Identity Management System System used to ensure that the right individuals have access to the right resources at the right times for the right reasons.	Complete	In progress	March 2017	January 2019	Yes

System name and description	**Status of development**	**Status of testing**	**Expected/ actual first deployment date[a] (delay since August 2017)**	**Expected/ actual final deployment date[a] (delay since August 2017)**	**Has at least a portion of the system's functionality been deployed?**
28. ECaSE –Field Operational Control System that manages field assignments, reviews and approves field worker's time and expense, and tracks field worker's performance.	Complete	In progress	July 2017	July 2018 (4-month delay)[b]	Yes
29. Geospatial Services Tool that provides vintage imagery service, internal current imagery service, public current imagery service, and mapping services.	Complete	In progress	July 2017	July 2018	Yes
30. Service Oriented Architecture Enterprise software architecture model used for designing and implementing communication between mutually interacting software applications in a service-oriented architecture.	Complete	In progress	July 2017	January 2019 (6-month delay)	Yes
31. ECaSE Operational Control System that manages the data collection universe for all enumeration operations, maintains operational workloads, and provides alerts to management.	Complete	In progress	August 2017	July 2018 (4-month delay)[b]	Yes
32. National Processing Center Printing Service that provides printing services for low-volume forms and merges static form and variable data, such as printing a standard form with unique addresses.	Complete	In progress	August 2017	July 2018	Yes

Table 3. (Continued)

System name and description	Status of development	Status of testing	Expected/ actual first deployment date[a] (delay since August 2017)	Expected/ actual final deployment date[a] (delay since August 2017)	Has at least a portion of the system's functionality been deployed?
33. ECaSE –Enumeration Tool that captures survey responses collected by door-to-door enumeration, records contact attempts, and collects employee availability and time and expenses.	Complete	In progress	March 2018 (1-month delay)	July 2018 (4-month delay)[b]	Yes
34. Integrated Computer Assisted Data Entry Tool that captures paper responses from questionnaires.	Complete	In progress	March 2018 (1-month delay)	July 2018	Yes
35. Census Image Retrieval Application that provides secure access to census data and digital images of the questionnaires from which the data were captured.	Complete	In progress	March 2018 (1-month delay)	January 2019 (11-month delay)[b]	Yes
36. Centurion Tool that provides an external interface for the upload of group quarters electronic response data.	Complete	In progress	July 2018 (4-month delay)	n/a	No
37. Sampling, Matching, Reviewing, and Coding System that is to support quality control for field operations.	In progress	In progress	August 2017	October 2018 (7-month delay)[b]	Yes
38. Control and Response Data System that is to provide a sample design and universe determination for the Decennial Census.	In progress	In progress	February 2018	January 2019 (3-month delay)	Yes

System name and description	Status of development	Status of testing	Expected/ actual first deployment date[a] (delay since August 2017)	Expected/ actual final deployment date[a] (delay since August 2017)	Has at least a portion of the system's functionality been deployed?
39. Census Data Lake Repository for response data that is to provide data access to reporting and analytics applications.	In progress	In progress	March 2018 (1-month delay)	January 2019	Yes
40. Decennial Response Processing System that is to perform data processing on the raw response data and stores the final processed response data for long-term storage.	In progress	In progress	March 2018 (1-month delay)	January 2019	Yes
41. Production Environment for Administrative Records Staging, Integration and Storage Tool that is to manage Administrative Records and provides services associated with those records.	In progress	In progress	March 2018 (1-month delay)	January 2019 (10-month delay)[b]	Yes
42. Fraud Detection System that is to identify fraudulent responses either in real-time or post data collection.	In progress	In progress	October 2018 (8-month delay)	n/a	No

Table 3. (Continued)

System name and description	Status of development	Status of testing	Expected/ actual first deployment date[a] (delay since August 2017)	Expected/ actual final deployment date[a] (delay since August 2017)	Has at least a portion of the system's functionality been deployed?
43. Center for Enterprise Dissemination Services and Consumer Innovation Tool that will provide search and access to tabulated Census data.	In progress	In progress	January 2019	n/a	No
44. Tabulation Tool that is to receive post-processed response data and produces tabulated statistical data.	In progress	In progress	January 2019	n/a	No

Source: GAO analysis of Census Bureau data. | GAO-18-655.

Legend: n/a = not applicable. These systems are only being deployed one time, so the first deployment date also represents the final deployment date.

[a] The dates listed for June 2018 or earlier should be considered actual dates.

[b] According to officials within the Bureau's 2020 Census Systems Engineering and Integration office, the delay in the final deployment date for this system is due to a change in the timing of the operations it is supporting for the 2018 End-to-End Test.

The Honorable Mark Meadows
Chairman

The Honorable Gerald E. Connolly
Ranking Member
Subcommittee on Government Operations
Committee on Oversight and Government Reform
House of Representatives

The Honorable Will Hurd
Chairman

The Honorable Robin L. Kelly
Ranking Member
Subcommittee on Information Technology
Committee on Oversight and Government Reform
House of Representatives

The Honorable Thomas R. Carper
United States Senate

APPENDIX I: STATUS OF DEVELOPMENT AND TESTING FOR SYSTEMS IN THE 2018 END-TO-END TEST, AS OF JUNE 2018

As part of the 2018 End-to-End Test, the Census Bureau (Bureau) plans to incrementally test, deploy, and use the 44 systems from December 2016 through the end of the test in April 2019. These operations include address canvassing, self-response (i.e., Internet, phone, or paper), field enumeration, and tabulation and dissemination. According to the Bureau, a single system may be deployed multiple times throughout the test (with additional or new functionality) if that system is needed for more than one of these operations.

Table 3 describes the status of development and testing, and describes if a portion of functionality has been deployed for each system in the 2018 End-to-End Test. The table also describes key system deployment dates and the delay in these dates since August 2017.

APPENDIX II: COMMENTS FROM THE DEPARTMENT OF COMMERCE

UNITED STATES DEPARTMENT OF COMMERCE
The Secretary of Commerce
Washington, D.C. 20230

August 17, 2018

Mr. David Powner
Director
Information Technology Management Issues
U.S. Government Accountability Office
441 G Street, NW
Washington, DC 20548

Dear Mr. Powner:

The U.S. Department of Commerce appreciates the opportunity to comment on the U.S. Government Accountability Office's (GAO) draft report, *"2020 Census: Continued Management Attention Needed to Address Challenges and Risks with Developing, Testing, and Securing IT Systems"* (GAO-18-655). We have no substantive disagreements with the findings in this report.

While we agree that much work remains to prepare and implement our information technology (IT) systems for the 2020 Census, we appreciate GAO's recognition of the substantial progress we have made over the past year as a result of increased management focus in the areas of cost estimation, scheduling, IT security, contract management, and governance. These include:

- An extensive and independent review at the Secretary's request of the lifecycle cost estimate for the entire 2020 Census, including all key assumptions, methods, and risk analyses for both IT and non-IT operations.
- Executing a detailed schedule development effort from July – December 2017 with the baselining of the schedule on December 14, 2017. The baselining has been followed by weekly statusing and monitoring of the schedule. Currently, we are working with Operations, Systems, and Testing Project Managers to incorporate the converted systems releases into the baselined Integrated Master Schedule by October 31, 2018.
- We completed security authorizations for all 43 systems currently operational and in use for the 2018 Census Test. The 44th and final system to be used in the 2018 Census Test is scheduled for authorization prior to its scheduled use in October of this year.
- We are continuously working with the Federal intelligence community and private-sector companies to strengthen our cybersecurity posture and to improve our incident response capabilities. We have implemented all Federal requirements and industry best practices, including use of both Federal and private-sector third parties to test our cyber defenses.

- Over the past year, we have increased our contract oversight and monitoring efforts with the addition of a weekly meeting where Decennial senior leadership meets to review contract performance status. Contract performance issues presented at this leadership status meeting are addressed to reduce risk to 2020 operations. In addition, we implemented monthly reporting on the 2020 Major Contracts using an EVM or EVM-like processes to track performance against cost and schedule, which are presented during a monthly status meeting with Secretary Ross.
- Revising 2020 Census governance to operationalize the processes as we transition from Research & Testing to Peak Operations. The revised approach provides clear decision escalation, reduces meeting burden, and improves communication of decisions to all stakeholders.

We have also enclosed our technical comments on this draft report.

Sincerely,

Wilbur Ross

Wilbur Ross

In: Top 5 Reports and Testimonies … ISBN: 978-1-53615-511-2
Editors: C. Johnston and W. Payne

Chapter 2

AIRCRAFT CARRIER DISMANTLEMENT AND DISPOSAL: OPTIONS WARRANT ADDITIONAL OVERSIGHT AND RAISE REGULATORY QUESTIONS*

United States Government Accountability Office

ABBREVIATIONS

ACAT	Acquisition category
DOD	Department of Defense
NRC	Nuclear Regulatory Commission
OMN	Operation and Maintenance, Navy
SCN	Shipbuilding and Conversion, Navy

* This is an edited, reformatted and augmented version of the United States Government Accountability Office Report to Congressional Committees, Publication No. GAO-18-523, dated August 2018.

Why GAO Did This Study

The Navy is planning to dismantle and dispose of CVN 65 after 51 years of service. In 2013, the estimated cost to complete the CVN 65 work as originally planned increased to well over $1 billion, leading the Navy to consider different dismantlement and disposal options.

The Senate Report accompanying a bill for the National Defense Authorization Act for Fiscal Year 2018 included a provision for GAO to review the Navy's plans for CVN 65. This chapter addresses (1) dismantlement and disposal options under consideration; (2) nuclear regulatory authority considerations; and (3) funding and reporting practices.

GAO reviewed budget, cost, and schedule documentation, as well as applicable laws, regulations, executive orders, policies, and guidance. GAO interviewed officials from the Navy and commercial companies about the dismantlement and disposal options, and NRC and state agencies about regulatory considerations.

What GAO Recommends

Congress should consider requiring Naval Reactors to coordinate with NRC to identify the applicable regulatory authority for a CVN 65 commercial dismantlement and disposal. GAO is also making four recommendations, including that the Navy take action to provide additional budget information and reporting to facilitate improved transparency and accountability for the CVN 65 cost, schedule, and risks. The Department of Defense agreed with all four recommendations.

What GAO Found

The Navy is assessing two options to dismantle and dispose of its first nuclear-powered aircraft carrier—ex-USS *Enterprise* (also known as CVN

65). CVN 65 dismantlement and disposal will set precedents for processes and oversight that may inform future aircraft carrier dismantlement decisions.

Characteristics of the Navy's Potential CVN 65 Dismantlement and Disposal Options

	Naval shipyard option	Full commercial option
General approach	Puget Sound Naval Shipyard dismantles a distinct section of the ship—the propulsion space section—that contains the 8 defueled reactors and all other nuclear-related material Shipyard prepares reactor packages for transport and disposal at facility in Hanford, Washington Commercial company dismantles and recycles or disposes of non-nuclear sections	Commercial company(ies) dismantles entire ship; potential companies and work locations yet to be determined Nuclear-related dismantlement uses applicable industry work practices—may include cutting into smaller components for shipping; disposal site(s) yet to be determined Commercial company recycles or disposes of non-nuclear portions
Navy preliminary cost estimate (dollars)	1.05 billion-1.55 billion	750 million-1.4 billion
Navy preliminary schedule estimate	10 years, 2034 start	About 5 years, 2024 start
Nuclear regulatory authority	Naval Reactors (Department of Energy)	Disagreement exists between Naval Reactors and Nuclear Regulatory Commission

Source: GAO analysis of Navy and Nuclear Regulatory Commission information. | GAO-18-523.

The Navy could rely on its extensive regulatory experience for the naval shipyard option. However, the Navy's ability to effectively evaluate the full commercial option is hampered by a disagreement with the Nuclear Regulatory Commission (NRC), which oversees the commercial nuclear industry. Naval Reactors officials assert that NRC's regulatory authority should apply to the full commercial option. NRC disagrees with this position. Coordination between the two agencies to identify the applicable regulatory authority and craft a regulatory plan would help ensure accountability, solidify cost estimates, and facilitate a CVN 65 decision.

The budget documentation and reporting that the Navy typically uses for ship dismantlement and disposal projects will not enable adequate oversight of CVN 65—a multi-year project with a cost that may exceed $1 billion. The documents that support Navy budget requests for dismantlement and disposal funding do not provide data that decision makers can readily use to track dismantlement costs against an established baseline or to evaluate funding plans for future years. Further, the Navy has no reporting requirements to support accountability for CVN 65 activities. Large defense acquisition programs generally are required to submit more detailed budget information and report on cost, schedule, and performance. These practices could be adapted for CVN 65 to provide information that will facilitate oversight commensurate with the scale of the effort.

August 2, 2018
Congressional Committees

After 51 years of service, ex-USS *Enterprise* (also known as CVN 65)— the Navy's first nuclear-powered aircraft carrier—is being prepared for dismantlement and disposal. At approximately 76,000 tons, CVN 65 will require an unprecedented level of work to dismantle and dispose of as compared to previous ships. The Navy originally intended to dismantle the entire CVN 65, both nuclear and non-nuclear components, at Puget Sound Naval Shipyard and Intermediate Maintenance Facility (hereafter referred to as Puget Sound Naval Shipyard), which is its usual facility for this type of activity. However, in 2013, the Navy's cost estimate for the shipyard to perform all CVN 65 dismantlement and disposal activities increased—from a range of $500 million to $750 million—to well over $1 billion. This led the Navy to consider alternatives. In 2016, the Navy issued a request for proposals to have a commercial company recycle the non-nuclear portions of the ship. The Navy also sought information from industry at that time on the potential for a commercial company to dismantle and dispose of the entire ship. In February 2017, the Navy announced it canceled its request for proposals on commercial recycling of non-nuclear portions of the ship and continued assessing its options.

As part of this assessment, the Navy determined it will not dismantle the entire ship at Puget Sound Naval Shipyard. Instead, it is focused on two potential options. One involves Puget Sound Naval Shipyard dismantling and disposing of a section of the ship that contains the nuclear material—referred to as the propulsion space section—with disposal at the Department of Energy's Hanford low-level radioactive waste disposal site in the state of Washington. The other entails the Navy awarding a contract for commercial industry to fully dismantle and dispose of the nuclear and non-nuclear materials for the entire ship. For the purposes of this chapter, we refer to these two options as (1) the naval shipyard option and (2) the full commercial option, respectively.

Regardless of the approach the Navy chooses, CVN 65 will set precedents for the processes, costs, and oversight that may be used to dismantle and dispose of nuclear-powered aircraft carriers in the future, such as the Nimitz-class carriers which the Navy will begin to retire in the mid-2020s. Senate Report 115-125 accompanying a bill for the National Defense Authorization Act for Fiscal Year 2018 included a provision for GAO to review the Navy's plans for CVN 65 dismantlement and disposal.

This chapter (1) describes the differences between the dismantlement and disposal options under consideration, including cost and schedule as well as workload and facilities; (2) evaluates the Navy's funding and reporting practices for dismantlement and disposal activities; and (3) assesses the effect that nuclear regulatory authority considerations have on dismantlement and disposal options for CVN 65.

To conduct our work, we reviewed documentation on prior, ongoing, and future Navy ship dismantlement and disposal plans and activities, as well as information specific to the different CVN 65 options the Navy has considered or is considering. We used this information to evaluate the Navy's history with ship dismantlement and disposal, how its plans for CVN 65 have evolved, and what is currently known about the different options the Navy is considering for the ship. We also reviewed the Navy's preliminary cost and schedule information and assessed naval shipyard workload and facility data. This included workload data for fiscal years 2006 through 2025 and facility data for fiscal years 2018 through 2035. To assess the reliability of these data, we interviewed knowledgeable officials and reviewed documentation to verify the controls and measures used to validate and maintain the data. We determined these data to be reliable for our purposes of discussing the existing and planned workload at Puget Sound Naval Shipyard. We also reviewed documentation on the Navy's funding and reporting practices, as well as regulatory authorities and processes for nuclear materials that may apply to CVN 65 dismantlement and disposal. This included a review of applicable laws, regulations, executive orders, policies, and guidance. Further, we assessed the typical budget information and reporting requirements for dismantlement and disposal in relation to federal internal control standards.[1] Additionally, we reviewed past GAO reports that addressed operation and maintenance activities at naval shipyards, and the processes, facilities, and requirements for nuclear materials.[2] We also reviewed federal and Department of Defense (DOD) budget, financial, and acquisition policies and guidance, as well as relevant reports from DOD and other nuclear energy-related organizations, such as the Organisation for Economic Co-operation and Development's Nuclear Energy Agency.

In addition to reviewing documentation that addresses these areas, we interviewed officials from the Naval Sea Systems Command—which includes Naval Reactors—Puget Sound Naval Shipyard, and the Nuclear Regulatory Commission (NRC). We also interviewed officials from health and environmental agencies for the states of Texas and Washington. Ship

---

[1] GAO, *Standards for Internal Control in the Federal Government*, GAO-14-704G (Washington, .C.: September 2014).

[2] GAO, *Naval Shipyards: Actions Needed to Improve Poor Conditions that Affect Operations*, GAO-17-548 (Washington, D.C.: Sept. 12, 2017); *Depot Maintenance: Executed Workload and Maintenance Operations at DOD Depots*, GAO-17-82R (Washington, D.C.: Feb. 3, 2017); *Nuclear Regulatory Commission: Regulatory Fee-Setting Calculations Need Greater Transparency*, GAO-17-232 (Washington, D.C.: Feb. 2, 2017); *Nuclear Security: NRC Has Enhanced the Controls of Dangerous Radioactive Materials, but Vulnerabilities Remain*, GAO-16-330 (Washington, D.C.: July 1, 2016); and *National Environmental Policy Act: Little Information Exists on NEPA Analyses*, GAO-14-370 (Washington, D.C.: Apr. 15, 2014)

dismantlement activities have occurred in these states and both have nuclear waste disposal sites. Additionally, we interviewed officials and reviewed documentation from commercial companies involved in shipbreaking or nuclear-related industries. These companies include Atkins Global; Energy Solutions; Huntington Ingalls Industries (HII – Nuclear); International Shipbreaking Limited; North Star Group Services; and Waste Control Specialists. For more information on our scope and methodology, see appendix I.

We conducted this performance audit from August 2017 to August 2018 in accordance with generally accepted government auditing standards. Those standards require that we plan and perform the audit to obtain sufficient, appropriate evidence to provide a reasonable basis for our findings and conclusions based on our audit objectives. We believe that the evidence obtained provides a reasonable basis for our findings and conclusions based on our audit objectives.

BACKGROUND

In 1961, the Navy commissioned the first and only Enterprise-class aircraft carrier, CVN 65, which was the world's first nuclear-powered aircraft carrier.[3] CVN 65 served the Navy's needs for 51 years, deploying 25 times and sailing more than 1 million miles during that time. The carrier, which was powered by eight nuclear reactors, was the predecessor of the two-reactor Nimitz-class aircraft carriers that followed it into service. The Navy plans to begin retiring the Nimitz-class carriers in the next decade.

[3] A nuclear-powered ship is constructed with the nuclear power plant(s) inside the engineering spaces. Each nuclear reactor onboard a U.S. Navy warship is located inside a specifically engineered space called the reactor compartment. A nuclear reactor is the heart of a nuclear power plant, in which nuclear fission may be initiated and controlled in a self-sustaining chain reaction to generate energy or produce heat. In shipboard nuclear reactors, the heat of nuclear reaction is transferred to water in a secondary steam system. The steam is then used as the source of power for the propulsion plant.

Following the retirement of CVN 65 in 2012, the Navy began preparing the ship for dismantlement and disposal in a process called inactivation. These inactivation activities—which Navy officials stated cost $863 million to complete—included removing the nuclear fuel from the ship's reactors and taking off equipment and other materials in preparation for dismantlement of the ship. The Navy's next steps include planning efforts to meet the environmental requirements associated with dismantling and disposing of a nuclear-powered ship, such as handling of radioactive and other hazardous materials. The final step for CVN 65 will be dismantlement, including the recycling of non-nuclear portions of the ship and safe disposal of nuclear and other hazardous materials. Figure 1 provides a timeline of CVN 65 events.

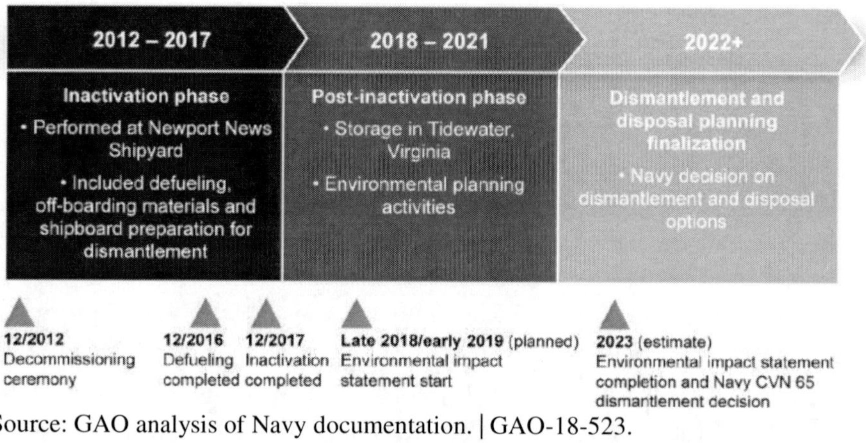

Source: GAO analysis of Navy documentation. | GAO-18-523.

Figure 1. CVN 65 Aircraft Carrier Timeline of Events Following Decommissioning.

CVN 65 is the largest nuclear-powered ship that has been retired by the Navy. Figure 2 compares the size of CVN 65 to previous and future Navy vessels requiring dismantlement and disposal, as well as other relatable structures.

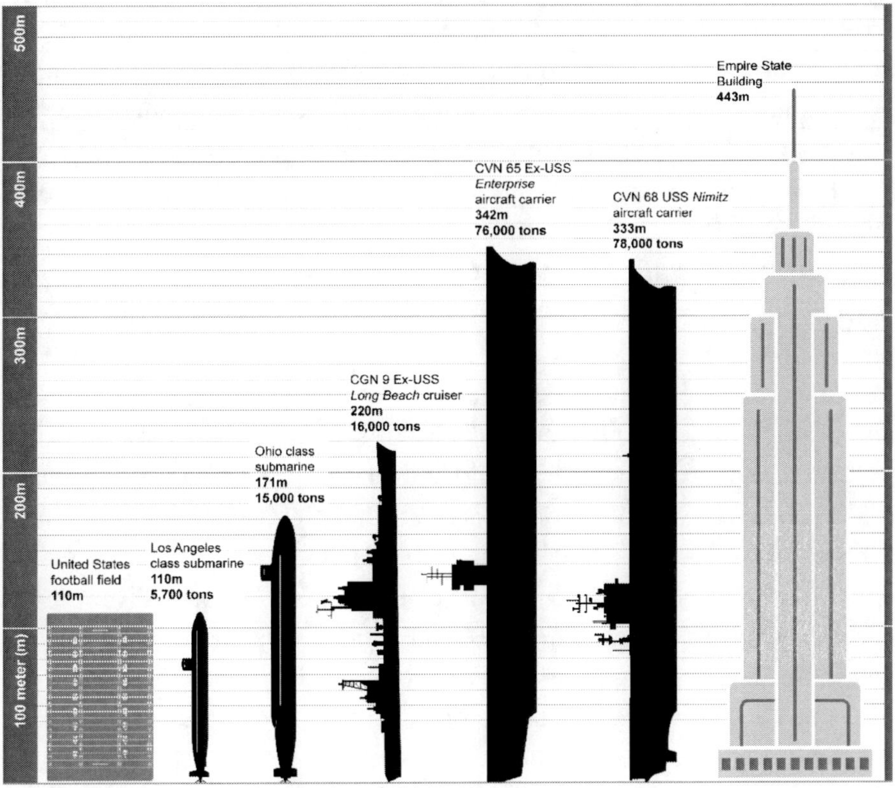

Source: GAO analysis of Navy data. | GAO-18-523.

Figure 2. Relative Size of CVN 65 Nuclear-Powered Aircraft Carrier.

Puget Sound Naval Shipyard Dismantlement and Disposal Activities

In 1990, the Navy authorized a program to recycle decommissioned submarines at Puget Sound Naval Shipyard in Bremerton, Washington. According to Navy officials, the Department of Energy's low-level waste site in Hanford, Washington, was the only practical site at the time for disposal of the defueled submarine reactor compartments, which included low-level radioactive waste.[4] Puget Sound Naval Shipyard is the largest shipyard on the U.S. West Coast, and while it is equipped and staffed to work on all classes of Navy vessels, it primarily conducts maintenance on nuclear-powered aircraft carriers and submarines, which the Navy considers a priority. This shipyard has the only dry dock on the West Coast capable of servicing an aircraft carrier and is the Navy's only site for dismantlement and disposal of nuclear-powered ships.

Since 1990, the Navy has inactivated over 130 nuclear-powered vessels. Inactivation is the process used to prepare a ship for disposing of the compartments that house the reactors and recycling the hull or for safe storage pending dismantlement and disposal at a later date. Inactivation includes draining hydraulic systems and tanks, and removing hazardous and expendable materials, tools, spare parts, and furnishings from the ship. The removal of the spent fuel from a ship's nuclear reactor(s), referred to as defueling, usually happens as part of inactivation. Historically, when a ship is dismantled at Puget Sound Naval Shipyard, the reactor compartments are removed and packaged for transport to the Hanford low-level radioactive waste disposal site. Figure 3 shows the typical path followed for dismantlement and disposal at the shipyard.

Dismantlement and Disposal by Commercial Industry

The Navy often uses commercial industry to dismantle and recycle its non-nuclear ships, including aircraft carriers, such as ex-USS *Constellation* and ex-USS *Ranger* completed in 2017. Navy officials noted that the cost to the government in recycling recent ships has been minimal—ranging from 1¢ to $6 million—because of the resale value of their scrap metal.

---

4 Low-level radioactive waste includes items that have become contaminated with radioactive material or have become radioactive through exposure to neutron radiation. It typically consists of contaminated protective shoe covers and clothing, reactor water treatment residues, and equipment and tools, among other items. Radioactivity can range from just above background levels found in nature, to very highly radioactive in certain cases, such as what may occur with parts from inside the reactor vessel in a nuclear power plant.

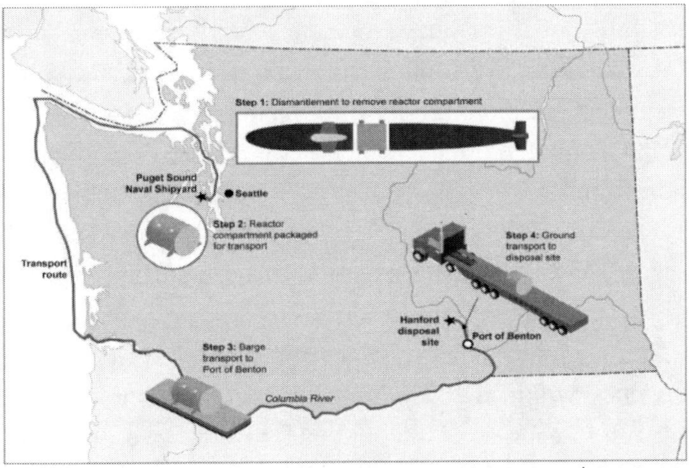

Source: GAO analysis of Navy data (data); Map Resources (map). | GAO-18-523.

Figure 3. General Process for Nuclear-Powered Ship Dismantlement and Disposal at Puget Sound Naval Shipyard.

Commercial companies have decommissioned 32 civilian nuclear reactor plants—work that the Navy has noted is comparable to nuclear-powered ship dismantlement and disposal. Commercial industry uses a component-based process for commercial nuclear plant decommissioning. This process breaks the reactor down into smaller components for transport and disposal, and separates nuclear waste from non-nuclear waste as much as possible to reduce disposal costs.

Requirements Related to Dismantlement and Disposal of Nuclear-Powered Ships

Several laws and an executive order have established the regulatory authority and requirements underlying the dismantlement and disposal of nuclear-powered Navy vessels. The Atomic Energy Commission exercised control of nuclear technology primarily for military purposes until 1954, when the Atomic Energy Act was amended.[5] These amendments allowed

[5] Atomic Energy Act of 1954, Pub. L. No. 83-703, 68 Stat. 919 (codified as amended at 42 U.S.C. §§ 2011-2297g-4 (2018)).

for the possibility of a privatized nuclear energy industry. Twenty years later, the Atomic Energy Commission was abolished and split into the Nuclear Regulatory Commission (NRC) and the Energy Research and Development Administration—which was later absorbed into the Department of Energy.

Under this structure, NRC is responsible for overseeing commercial nuclear reactor safety, licensing reactors, and establishing regulations and guidelines for radioactive waste disposal for the commercial nuclear industry. The National Nuclear Security Administration, a separately organized agency within the Department of Energy, is responsible for the management and security of the nation's nuclear weapons, as well as nonproliferation programs. The Naval Nuclear Propulsion Program—also known as Naval Reactors—is a joint program of the Department of Energy and DOD that has cradle-to-grave responsibility for all naval nuclear propulsion matters. Figure 4 provides a brief description of laws and orders related to nuclear materials.

In addition to the nuclear-specific requirements guiding the dismantlement and disposal process, the Navy must comply with the National Environmental Policy Act.[6] Specifically, this act requires federal agencies to evaluate the likely environmental effects of projects they are proposing, generally by preparing either an environmental assessment or a more detailed environmental impact statement.[7] An environmental impact statement must, among other things, (1) describe the environment that will be affected, (2) identify alternatives to the proposed action and identify the agency's preferred alternative, (3) present the environmental impacts of the proposed action and alternatives, and (4) identify any adverse

[6] 42 U.S.C. §§ 4321-4347 (2018).

[7] Agencies may prepare an environmental assessment to determine whether a proposed project is expected to have a potentially significant impact on the human environment. If the agency determines that the project may cause significant environmental impacts, an environmental impact statement should be prepared. However, if the agency determines there are no significant impacts from the proposed project, then it is to prepare a Finding of No Significant Impact that presents the reasons why the agency has reached that conclusion. If a proposed project fits within a category of activities that an agency has already determined normally does not have the potential for significant environmental impacts or the activity has been excluded from National Environmental Policy Act analysis by statute, then the agency may instead approve it by using a categorical exclusion.

environmental impacts that cannot be avoided should the proposed action be implemented. The Act's requirements are invoked for major federal actions, such as the construction of buildings or highways, or the dismantlement and disposal of reactor compartments from nuclear-powered vessels.

Atomic Energy Act of 1946	Atomic Energy Act of 1954	Energy Reorganization Act of 1974	Executive order 12344 Issued in 1982[a]	Low-Level Radioactive Waste Policy Amendments Act of 1985, as amended
Emphasized military aspects of nuclear energy and did not provide for commercial application of atomic energy Established the Atomic Energy Commission	Authorized the Atomic Energy Commission to continue its weapons program, promote commercial uses of nuclear power, and protect against hazards related to the possession and use of nuclear materials	Abolished the Atomic Energy Commission and divided its functions between the Energy Research and Development Administration (later part of the Department of Energy) and the Nuclear Regulatory Commission	Sets forth Naval Reactors' responsibility for all aspects of the Navy's nuclear propulsion, including research, design, construction, testing, operation, maintenance, and ultimate disposition of naval nuclear propulsion plants	Gives states the responsibility to dispose of low-level radioactive waste generated within their borders and allows them to form compacts to locate facilities to serve a group of states and the facilities will be regulated by the Nuclear Regulatory Commission or by states

Source: GAO analysis of relevant laws and an exclusive order. | GAO-18-523.
[a] See also 50 U.S.C. §§ 2406, 2511 (2018).

Figure 4. Laws and Executive Order Related to Nuclear Materials.

Since 1996, nuclear-related dismantlement and disposal activities performed by Puget Sound Naval Shipyard have been based on the same environmental impact statement—which addresses the effects of disposing of submarine and cruiser reactor compartments. In 2012, the Navy produced an environmental assessment analyzing the effects of removing and preparing the reactor compartments of CVN 65 for disposal at Puget Sound Naval Shipyard and transporting the compartment packages to the Hanford site for disposal. It found that these activities would have no significant impact on the environment beyond existing activity. Naval Reactors subsequently decided, however, that a new environmental impact statement is required for CVN 65 because the alternatives identified for dismantling and disposing of the ship could potentially have significant impacts on the environment that are not captured by the existing environmental assessment. As part of the new statement for CVN 65, Navy officials said environmental factors that account for the naval shipyard and full commercial options will be reviewed, as well as indefinite waterborne storage of the ship pending dismantlement and disposal at a later date.

NAVAL SHIPYARD OPTION FOR CVN 65 IS MORE DEFINED THAN FULL COMMERCIAL OPTION BUT MAY POSE CHALLENGES FOR MEETING NAVY PRIORITIES

The Navy is weighing a number of considerations before making a decision for CVN 65 dismantlement and disposal. The naval shipyard option offers well-established processes for dismantlement and disposal of the ship's nuclear material and better understood cost and schedule estimates than the full commercial option. Our analysis of available data, however, found that the naval shipyard option would contribute to existing workload backlogs and exacerbate facility challenges at the shipyard that could affect its work maintaining the active fleet—a Navy priority. While the Navy has not defined its requirements for the full commercial option, industry does not expect to face workload or facility challenges. Navy officials also believe that the full commercial option potentially could shorten the timeline for completing the work and reduce the total cost.

Naval Shipyard Option Is Based on a Well-Established Process, While Navy Has Yet to Characterize Full Commercial Option

Although CVN 65 is the first nuclear-powered aircraft carrier requiring dismantlement and disposal, the Navy has well-established processes for dismantling and disposing of nuclear-powered submarines and cruisers. Navy officials explained that the shipyard's extensive dismantlement and disposal experience with these vessels has resulted in a strong understanding of how to accomplish the work. Further, the Navy has been working on plans to address the ship-specific needs of CVN 65 for many years. If the Navy chooses the naval shipyard option for CVN 65, it expects to adapt and use these well-established processes to dismantle the 28,000-ton nuclear propulsion space section at Puget Sound Naval Shipyard. This section would contain the 8 defueled reactors and all other nuclear-related material that remains on the ship. To separate the propulsion space from the ship, a commercial company would perform "ship-shaping" to create a dedicated ship section for all of the nuclear-related work. This activity would minimize the portion of the ship transported to the naval shipyard for dismantlement and disposal. The remaining ship sections would be commercially recycled. The shipyard is evaluating two designs for reactor compartment packages that could be used for transport and disposal of the ship's nuclear material. One design—based on a package previously used for cruiser reactors—would involve the shipyard preparing 8 single reactor packages. The other includes a new design that would enclose 2 reactors in dual reactor packages. Figure 5 shows how the Navy anticipates the ship would be divided into sections through this process.

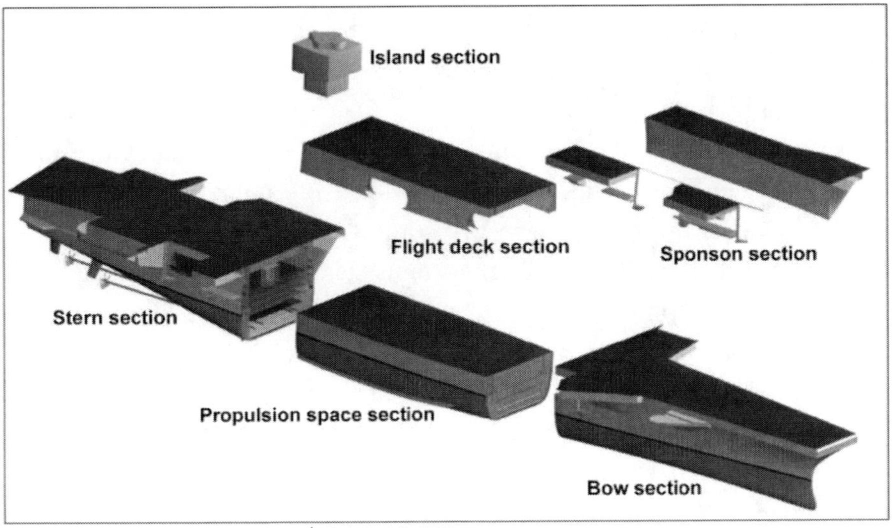

Source: Navy documentation. | GAO-18-523.

Figure 5. Proposed CVN 65 Sectioning Approach for Puget Sound Naval Shipyard Dismantlement Option.

In contrast, the Navy formally began considering the potential for a full commercial option for CVN 65 within the past 4 years. According to Navy officials, although information received through previous requests for information and hosting discussions with commercial industry helped shape their understanding of the potential for a commercial ship dismantlement, they ultimately have had relatively limited interaction with commercial companies to determine their potential plans and processes for CVN 65 dismantlement and disposal. Naval Reactors officials stated they are waiting for the environmental impact statement process to officially begin before further engaging with prospective commercial companies and the public.

Many of the details for a full commercial option will depend upon Navy requirements, such as standards, technology, or specific procedures required to do the work; data and analysis in the environmental impact statement; and preferred work practices and facilities of prospective companies. Officials we interviewed from companies with potential interest in the work stated that because the Navy has not communicated its CVN 65 requirements for a full commercial option, any commercial approach described for the work would be hypothetical at this point, relying on their extensive prior experience with nuclear materials handling, packaging, shipping, and disposal—including nuclear ship In contrast, the Navy formally began considering the potential for a full commercial option for CVN 65 within the past 4 years. According to Navy officials, although information received through previous requests for information and hosting discussions with commercial industry helped shape their understanding of the potential for a commercial ship dismantlement, they ultimately have had relatively limited interaction with commercial companies to determine their potential plans and processes for CVN 65 dismantlement and disposal. Naval Reactors officials stated they are waiting for the environmental impact statement process to officially begin before further engaging with prospective commercial companies and the public.

Table 1. Characteristics of the Navy's Potential CVN 65 Dismantlement and Disposal Options

	Naval shipyard option	**Full commercial option**
General approach	Commercial company divides CVN 65 into sections and creates a distinct nuclear propulsion space section in the middle of the ship that contains the 8 defueled reactors and all other nuclear-related material Company dismantles and recycles or disposes of non-nuclear sections Propulsion space section is transported by heavy-lift towing to Puget Sound Naval Shipyard Naval shipyard dismantles nuclear propulsion space section using established shipyard processes and constructs 4 or 8 reactor packages for transport and disposal	Commercial company(ies)dismantles entire shipbased on applicable industry practices; may be similar to component-based dismantlement used for nuclear power plant decommissioning Company recycles or disposes of non-nuclear portions
Nuclear-related dismantlement location	Puget Sound Naval Shipyard	Not yet determined
Disposal path	Transport reactor compartment packages by barge and ground transport vehicle to low-level radioactive waste disposal facility in Hanford, Washington	Not yet determined
Navy preliminary cost and schedule estimates	$1.05 billion-$1.55 billion 10 years, 2034 start	$750 million-$1.4 billion About 5 years, 2024 start

Source: GAO analysis of Navy documentation and interviews. | GAO-18-523.

Many of the details for a full commercial option will depend upon Navy requirements, such as standards, technology, or specific procedures required to do the work; data and analysis in the environmental impact statement; and preferred work practices and facilities of prospective companies. Officials we interviewed from companies with potential interest in the work stated that because the Navy has not communicated its CVN 65 requirements for a full commercial option, any commercial approach described for the work would be hypothetical at this point, relying on their extensive prior experience with nuclear materials handling, packaging, shipping, and disposal—including nuclear ship maintenance and decommissioning of commercial reactors—or ship recycling. Commercial company officials noted that despite the lack of definitive information available, they would anticipate employing typical practices used for commercial nuclear reactor decommissioning, ship dismantlement, and control of nuclear materials to complete CVN 65 work. In terms of locations for the work, Naval Reactors officials noted that many coastal sites in the United States could potentially accommodate CVN 65 dismantlement activities, and the location of the work site would affect the proposed disposal site or sites. Table 1 provides characteristics of the two options that the Navy is considering for CVN 65 dismantlement and disposal.

Estimates for Both Dismantlement Options Require Further Development

Cost and schedule estimates for both CVN 65 options have yet to be formally established by the Navy. Puget Sound Naval Shipyard has been refining CVN 65 plans and estimates over many years. However, its most recent estimates for cost and schedule still may not fully account for uncertainties in completing the work because it represents a first-of-its kind project with an unprecedented scale. The Navy's notional estimates for the commercial option are a first step in establishing expectations and will evolve as requirements for the work are better understood. The Navy awarded a contract in July 2018 to the Center for Naval Analyses—a federally funded research and development center serving the Navy and other defense agencies—to complete a cost analysis for the full commercial option. This effort is expected to provide the Navy with a cost estimate for CVN 65 in October 2018, followed by a model through which the Navy can develop cost estimates for future Nimitz-class dismantlement and disposal efforts. The findings from the CVN 65 environmental impact statement may contribute to the final cost and schedule estimates for either option.

Better Fidelity in Existing Naval Shipyard Option Estimates

Puget Sound Naval Shipyard officials explained that as their planning has progressed, they have refined their cost and schedule estimates for CVN 65 dismantlement and disposal. Overall, the Navy's cost estimates have increased significantly from initial estimates but have been relatively stable since 2016. The schedule went through similar fluctuations but has steadied. Table 2 outlines changes in the shipyard's plans and how they affected cost and schedule.

The schedule for starting the work at the naval shipyard also changed. Navy officials stated that as a result of the Navy's decision in early 2017 to reassess its options for CVN 65, it delayed the expected start date for the naval shipyard option from 2019 to 2034 based on analysis of the workload at the naval shipyard, which we discuss below.

Although Puget Sound Naval Shipyard officials noted their cost estimate includes some margin to account for CVN 65 being the first project of its kind, it may not adequately account for the extent of unknown facts or circumstances that could affect cost. For example, unrecognized hazardous materials may exist in inaccessible areas of the CVN 65 propulsion space section that will only be discovered once the work is underway, which could affect cost and schedule. Execution of the work in support of a new dual reactor compartment package design also could lead to unanticipated challenges that cause deviations from estimates.

No Formal Estimates for Full Commercial Option

The Navy has notionally estimated cost and schedule for a full commercial option to be $750 million to $1.4 billion and about 5 years to complete. These estimates suggest that the commercial option could cost less and take less time to complete than the naval shipyard option. Navy officials stated that the notional cost estimate is derived from data reported by nuclear power plant operators, with differences in size and scope for the nuclear reactors incorporated.[8] They also said that the notional estimate will be updated once it receives additional information from industry during the planning process.

Table 2. Changes in Puget Sound Naval Shipyard CVN 65 Plans, Estimate Methodology, Schedule, and Cost

Year	Planning history	Estimate methodology	Schedule duration	Cost (in then-year dollars)
2011	Puget Sound Naval Shipyard dismantles and disposes of entire ship beginning in 2017 - prepare 8 single reactor compartment packages - recycle non-nuclear portions	Based on workdays per ton to reflect returns of completed cruiser reactor packages	7 years	500 million-750 million
2013	Same approach as 2011	Developed manning profiles for each phase of work - added cost for removal of hazardous materials - added support costs for increased duration of effort	12 years, 9 months	1.25 billion-1.85 billion
2014	Puget Sound Naval Shipyard begins dismantlement and disposal of propulsion space section of ship in 2019 Commercial recycling of non-nuclear portions	Updated manning profiles - reduced non-nuclear work and duration at shipyard	12 years	950 million-1.4 billion

[8] The Nuclear Regulatory Commission noted these reported data are not equivalent to the actual cost to decommission a reactor, but represent the majority of the cost associated with safely decommissioning a single nuclear facility.

Year	Planning history	Estimate methodology	Schedule duration	Cost (in then-year dollars)
2016	Same overall approach as 2014 - prepare 4 dual reactor compartment packages	Based on actual ship checks, drawings, and work teams - included nuclear and fire safety oversight - added oversight for ship-shaping to create propulsion space section	10 years	1.05 billion-1.55 billion[a]

Source: GAO analysis of Navy documentation. | GAO-18-523.

Note: Navy officials stated that all estimates include an average of $113 million in capital investments to Puget Sound Naval Shipyard required to execute the CVN 65 reactor compartment package disposal, while also ensuring that doing so would not affect the shipyard's maintenance of active ships.

[a] Puget Sound Naval Shipyard included a $51 million (6 percent of labor costs) management reserve for risk mitigation.

Navy officials told us they expect the cost per reactor for CVN 65 would be significantly less than the NRC decommissioning average for a commercial facility because CVN 65 reactor compartments are smaller, the reactors are more compact, and they have already been through the costly defueling activity. A 2016 international study on the cost of decommissioning nuclear power plants identified several high-level categories and their contribution to total costs for reactors decommissioned in the United States, such as project management, site restoration, and waste packaging, transportation, and disposal.[9] According to this study, about 25 percent of decommissioning costs can be attributed to reactor decontamination and dismantling. Using this percentage and the average cost to decommission a commercial nuclear reactor, we estimate the cost to dismantle the eight CVN 65 defueled reactors to range from $1.2 billion to 1.3 billion, which is at the higher end of the Navy's notional estimated range for the full commercial option.

[9] Organisation for Economic Co-operation and Development, Nuclear Energy Agency, *Costs of Decommissioning Nuclear Power Plants,* NEA report No. 70201, 2016.

In addition to the potential cost, the Navy initially projected about a 5-year period of performance for the full commercial option based on limited industry input. Navy officials told us the full commercial option start date, beginning no earlier than 2024, is contingent on the finalization of the environmental impact statement and a record of decision that chooses this option as the Navy's path forward. The Navy's intent would be to award a contract shortly after the environmental impact statement is completed if the Navy decides to pursue the full commercial option. Commercial officials told us they do not anticipate a need for significant lead time before starting work, though the need will be better understood once the Navy outlines requirements for the work.

Finally, the cost for a full commercial option could be influenced by the contract type selected by the Navy. Contract type selection is a key factor in determining how cost risk is shared between the Navy and the contractor. Firm-fixed-price contracts are suitable for situations where the risk involved is minimal or can be predicted with an acceptable degree of certainty. Conversely, cost-type contracts are used when either requirements are not sufficiently defined or uncertainties with contract performance do not permit costs to be sufficiently estimated to use a fixed-price contract. Although no decision has been made, Navy officials told us they are interested in using a firm-fixed-price contract—a contract type that has been used for commercial reactor decommissioning. Under a firm-fixed-price contract, the contractor agrees to perform the work for a price that is not subject to change based on the contractor's cost experience in performing the contract, placing full responsibility for all costs and resulting profit on the contractor. Navy officials stated that because CVN 65 is the first nuclear-powered aircraft carrier to be disposed of, the scope of the effort will need to be better defined before they could reliably conclude that firm-fixed-price contracting would be appropriate. Specifically, insufficiently understood risks may make potential contractors unwilling to accept the risks associated with a firmfixed-price contract.

The Navy's Priorities for Puget Sound Naval Shipyard Present Challenges Not Expected for Full Commercial Option

The Navy has stated its priority for Puget Sound Naval Shipyard is the work associated with maintaining nuclear-powered aircraft carriers and submarines currently in the fleet. However, as we reported in 2017, Puget Sound Naval Shipyard has had significant fleet maintenance delays since fiscal year 2000.[10] These delays resulted in 4,720 lost operational days for nuclear-powered aircraft carriers and submarines. The addition of CVN 65 would contribute to challenges in the naval shipyard's ability to meet workload demands and further constrain its available facilities. In comparison, despite the lack of detail about the Navy's requirements, commercial company officials we interviewed stated they currently do not anticipate any major workload challenges or conflicts with other ongoing or future work in completing the work on CVN 65 based on their existing workforce and potential facilities for performing the work.

Puget Sound Naval Shipyard Workload and Facility Challenges

Based on our analysis of workload and resources data from Puget Sound Naval Shipyard, we found that the shipyard consistently operates at its maximum annual workload level and this likely will continue regardless of the Navy's decision for CVN 65. A Naval Reactors analysis of the shipyard's workload data also shows the workload meeting or exceeding capacity for the foreseeable future. The shipyard's workload projections that we reviewed show it will be working at or near capacity through fiscal year 2025—the last year for which data were available. Adding the work associated with dismantlement and disposal of CVN 65 would put the shipyard over current workload capacity.

Shipyard officials explained that historically, the workload projection for a given year matures as that year approaches, and the dips that sometimes are depicted in future-year workload projections generally vanish. Workload maturity or growth can be attributed to changes in the

---

[10] GAO-17-548.

Navy's maintenance plans, deferred maintenance, growth from the previous year, and overall shipyard productivity. The condition of a ship when it arrives for maintenance can also contribute to growth if inspections of systems or components reveal a need for unplanned repairs. To account for historical variability and improve projections of overall workload, in 2015 shipyard officials began including 10 percent in unallocated workload to projections.

In reviewing the shipyard workload and resources data, we also found that the shipyard regularly underestimates workload for future years—especially 5 years or more out—with workload growth for future years consistently exceeding 15 percent. Even without the CVN 65 work at the naval shipyard, projections show its workload with average notional growth will meet or exceed the workforce available to complete the work, as shown in figure 6.

According to the Navy, it is typical for naval shipyards to continually shift resources across projects to align worker-specific trade skills to the type of work executed on any hull in the shipyard, at any particular time. To achieve a level and sustainable workforce across the fiscal years, the number of full-time employees required to support planned work is sized as part of the total workforce. The shipyard mitigates peaks in workload (above the available workforce) through the use of additional overtime, loans from other naval shipyards, and contracting. When that cannot occur, the shipyard will defer workload until it can be executed.

The CVN 65 dismantlement and disposal work could affect the shipyard's ability to complete active fleet maintenance. We found that the addition of the CVN 65 dismantlement and disposal would add almost a year's worth of work across the estimated 10-year dismantlement and disposal period to an already busy shipyard that has demonstrated difficulties in accurately projecting its future work. The Navy prioritizes maintenance of the active fleet, but the scale of the CVN 65 work would reduce the shipyard's ability to delay or reprioritize dismantlement and disposal. Shipyard officials noted that the Navy often defers planned dismantlement and disposal to address higher-priority active fleet maintenance. For example, smaller submarines prepared for dismantlement can instead be stored at the shipyard until workforce and space are available to complete the work. However, an aircraft carrier—even when reduced to a propulsion space section as proposed for CVN 65—would not offer the same level of flexibility to defer work. CVN 65 would involve a more extensive resource commitment because of its increased size relative to past ship dismantlement projects and would occupy limited facilities at the shipyard. Specifically, current plans require 3 years pier side to prepare the propulsion space section for dismantlement and reactor compartment disposal and about 5 years in a dry dock for the actual dismantlement.

Further, the shipyard expects a significant increase in its submarine inactivation and reactor compartment disposal and hull recycling workload due to the end of service for an additional class of submarines—specifically, the Ohio-class submarines starting in 2027. The estimated increase in inactivation and reactor recycling workload would overlap with the planned start for CVN 65 dismantlement and disposal in 2034, if the Navy elects to pursue this option. In addition, the shipyard already has a backlog of 10 submarines and the ex-USS *Long Beach* cruiser in storage awaiting disposal and recycling at its long-term storage facility for defueled, decommissioned, and inactivated nuclear-powered ships. Another 3 submarines are pier-side at Puget Sound Naval Shipyard. This backlog is not expected to subside as submarines continue to be retired, and each vessel represents thousands of workdays that the shipyard has to commit to its dismantlement and disposal.

Navy and Industry Expect Full Commercial Option to Face Fewer Challenges

While the Navy has not established specific requirements for the full commercial option, Navy officials maintain that it does not present the same workload and facility challenges that exist for the naval shipyard option. Commercial companies have flexibility in selecting a location for CVN 65 dismantlement activities based on facility and workforce availability considerations. Some company officials we spoke with also noted they have existing worksites—which are audited and approved by Naval Reactors—where they process, package, and transport low-level radioactive waste or operate low-level radioactive waste disposal sites licensed by NRC. These include facilities for radioactive waste processing and decontamination of materials for recycling. Additionally, company officials said they anticipate that a substantial amount of the work could be performed with the ship in the water—similar to the traditional approach used to dismantle non-nuclear vessels for recycling—and existing contractor facilities likely would not require major upgrades or improvements other than to provide for the radiological-based waste handling and packaging considerations.

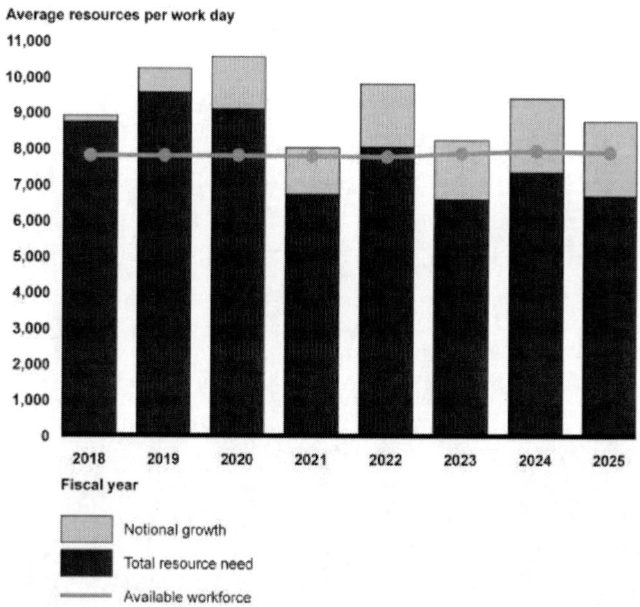

Source: GAO analysis of navy data. | GAO-18-523.

Notes: The average total resource need indicates the amount of work days that are estimated to complete planned work for each work day. When this exceeds the available workforce, the shipyard must use measures, such as overtime hours, to address the additional need. CVN 65 dismantlement and disposal workload projections are not included in these data.

Figure 6. Puget Sound Naval Shipyard Average Workload Projections with Available Workforce and Notional Growth without CVN 65 Work.

Commercial company officials told us that they would not expect significant additional hiring needs based on their limited understanding of the potential CVN 65 work and their existing workforce capacity. They added that the nuclear dismantlement and disposal industry has an available, qualified workforce that could easily be employed if additional workforce were needed. Given the early stage of the Navy's planning for CVN 65 and the Navy's lack of formal engagement with commercial companies at the time of our review, we did not assess the current or future commercial workforce capacity. Any details on potential CVN 65 facility and workforce plans from commercial companies will be hypothetical until the Navy formally begins efforts to seek input from commercial companies and communicate requirements.

Budget Documentation and Reporting Does Not Include Sufficient Information to Facilitate Transparency and Oversight for CVN 65

The Navy's approach typically used to budget for and report on ship dismantlement and disposal does not provide sufficient information to support decision makers' oversight of CVN 65—a multi-year project that may require more than $1 billion to complete. We found the Navy is not required to provide detailed budget information or report dismantlement and disposal cost, schedule, and programmatic information to decision makers. Providing additional information through budget requests and reporting would help ensure that decision makers have sufficient information to oversee CVN 65 dismantlement and disposal activities and to support future decisions.

Budget Exhibits for Dismantlement and Disposal Lack Ship-Specific Details

The Navy uses budget exhibits to provide congressional decision makers information about dismantlement and disposal efforts. If no changes are made to the information provided within the Operation and Maintenance, Navy (OMN) budget exhibits, the CVN 65 dismantlement and disposal budget request will include limited details for planned work, funding needs, and total estimated costs. The bulk of the Navy's past dismantlement and disposal work is comprised of comparatively low-cost projects—particularly submarines—with limited resource demands compared to a nuclear-powered aircraft carrier like CVN 65, a multi-year project with a cost that will potentially exceed $1 billion. For example, nuclear-powered submarines have an average dismantlement and disposal cost of about $26 million and average about 50,000 workdays. Federal internal control standards recommend that agency management communicate with external stakeholders the necessary quality information—such as complete cost and schedule information for CVN 65 dismantlement and disposal—to achieve objectives.[11] Budget exhibits are a primary source of information about all programs and other activities during budget planning and congressional appropriation decisions. Well-prepared budget exhibits help provide a rationale for the amount and timing of funding requests. Given that this multi-year, large-scale project is the first of its kind, more detailed information would facilitate greater transparency and oversight of cost, schedule, and performance.

Limited Budget Information Provided for Dismantlement and Disposal

The Navy uses the OMN appropriation account to fund dismantlement and disposal activities.[12] The Navy's Financial Management Policy Manual provides overall summary guidance on OMN budget formulation, but it does not provide specific guidance on reporting criteria for dismantlement and disposal of Navy ships.[13] Budget exhibits are prepared to justify appropriation requests and are key documents that can be used to support

[11] GAO-14-704G.

[12] OMN appropriations are for expenses not otherwise provided for, such as day-to-day costs of operating naval forces, including fuel, supplies, and maintenance of ships. Department of the Navy, Office of the Assistant Secretary of the Navy (Financial Management and Comptroller), Financial Management Policy Manual, Oct. 2017.

[13] Department of the Navy, Office of the Assistant Secretary of the Navy (Financial Management and Comptroller), Financial Management Policy Manual, Oct. 2017.

congressional oversight. DOD acquisition training materials state that well prepared budget exhibits make programs more defensible. However, in assessing the OMN budget exhibits associated with dismantlement activities for fiscal years 2007-2018, we found they provide little ship-specific detail that could be used to monitor a significant project such as the planned effort for CVN 65 dismantlement and disposal, which may begin requesting funding as soon as fiscal year 2023.

Specifically, we reviewed the dismantlement and disposal funding requests from the past several years, which reside within the Navy's OMN budget exhibits under the *Ship Activations/Inactivations* sub-activity group of the *Mobilization* budget activity. In doing so, we found these exhibits generally contain high-level information with a summary of funding changes for the current fiscal year and the requested funding estimate for the budget year. We could not definitively identify or track dismantlement and disposal of specific ships because

- key work activities are not described by ship,
- cost and schedule for individual ships are not presented, and
- prior year costs and cost to complete a specific ship's dismantlement and disposal are not provided.

In reviewing programmatic documentation other than the budget requests, such as Puget Sound Naval Shipyard dismantlement planning documents and the Navy's long-range shipbuilding plans, we found instances of submarine inactivation costs significantly exceeding estimates and notable delays to the start dates for work activities. We found that, although not required, this information was not reflected in the budget exhibit documents we reviewed. As another example we previously noted, Navy officials stated that CVN 65 inactivation—already completed in December 2017—cost $863 million. We could not track this cost from the budget exhibits because of their limited detail. As a consequence of the general lack of detail in the budget exhibits, decision makers cannot readily identify if cost growth occurred or if a specific ship was dismantled when planned, hindering oversight of dismantlement and disposal projects.

The Navy's OMN annual appropriations fund work activities on a year-by-year basis, which does not necessarily allow for tracking of the full resource commitment of a project over time or enable monitoring of cost growth to determine if additional funds are needed. Navy officials stated that they fully fund dismantlement and disposal efforts that span multiple fiscal years. They added that for CVN 65, the Navy may divide the work into multiple discrete phases that are separately funded due to the lengthy projected schedule. This approach could require the Navy to seek OMN appropriations in several non-consecutive years. Such an approach could make tracking CVN 65 dismantlement and disposal funding challenging, as the total cost and any changes would be obscured among the multiple funded activities that collectively compose the total dismantlement and disposal effort. Navy officials acknowledged that they could provide further information, such as total project cost and an overall schedule for CVN 65, in the OMN budget exhibits. However, without direction from DOD leadership or Congress, Navy officials stated that they have no plans to deviate from providing the traditional OMN budget exhibit information. Providing additional information in the CVN 65 budget exhibit could enable decision makers to track total cost, any cost changes, schedule progress, and general performance for the CVN 65 dismantlement and disposal.

Navy Could Provide More Budget Details for CVN 65

While the Navy funds ship dismantlement and disposal from the OMN account, budget exhibits for other accounts—such as the Shipbuilding and Conversion, Navy (SCN) account typically used for major investment items—offer examples of how to provide decision makers with more detailed information. Budget exhibits for SCN appropriations are structured to identify major elements of cost and track those costs over time, consistent with DOD Financial Management Regulations. For example, the SCN budget exhibits typically contain specific information for each ship being procured with a distinct funding line for major cost categories such as basic construction, propulsion, and electronics. Additionally, these budget exhibits describe the program with specific plans for the upcoming budget year and estimate across 5 fiscal years (known as DOD's Future Years Defense Program), including the total cost to complete the program.[14] While some of the SCN budget exhibit elements are not applicable to dismantlement and disposal, others could be adapted and used in an OMN budget exhibit for CVN 65 to provide information that would enable better oversight, such as

- work activities planned and performed by fiscal year;
- prior years' funding data;
- future years' funding plans;
- total estimated cost;
- cost to complete dismantlement;
- schedule of key events; and
- information on the contractor(s), contract type, and contract award and completion dates.

Navy officials said they typically would not provide the level of detail found in SCN budget exhibits or the exhibits for other DOD acquisition programs because OMN exhibits are not designed to support the same level of oversight. Unlike DOD acquisition programs, DOD projects completed with operation and maintenance funds typically are not investment programs and generally do not require the same level of oversight. However, as we previously indicated, Navy officials noted that if DOD leadership or Congress provided clear direction on what additional details related to CVN 65 dismantlement and disposal should be included in OMN budget exhibits, it could provide that additional information to support oversight. Navy officials stated that given the considerable funding needs and congressional interest with CVN 65, they were assessing options for providing specific detail in the OMN budget exhibit for its

[14] The Future Years Defense Program is the program and financial plan for DOD as approved by the Secretary of Defense. It arrays programmed dollars, manpower and force structure over a 5-year period (force structure for an additional 3 years), portraying this data by Major Force Program (i.e., strategic forces, mobility forces, and research and development) for DOD internal review for the program and budget review submission. It is also provided to the Congress in conjunction with the President's budget submission.

dismantlement and disposal activities. They added that no specific decisions had been made on what additional information, if any, would be included for CVN 65.

Lack of Reporting Requirements Limits Opportunities for Insight into CVN 65 Dismantlement and Disposal Cost, Schedule, and Performance

Despite being a part of the final phase in the program's life cycle, we found no specific reporting requirement related to the cost, schedule, risk management, and general performance of dismantlement and disposal activities in DOD or Navy policy that would support oversight by DOD or Congress.[15] Officials from Naval Reactors and the Naval Sea Systems Command confirmed that there is no reporting requirement for performance of dismantlement and disposal of Navy ships. Navy officials noted that dismantlement and disposal activities are included in their annual briefings to Congress that support the Navy's budget requests, but acknowledged that the typical comparatively low-cost ship dismantlement and disposal activities are generally of less interest when combined with a briefing on shipbuilding and other high-dollar acquisition investments. This approach may be appropriate for submarine dismantlement and disposal activities that have lower costs, shorter periods of performance, and a well-established history.

---

[15] DOD Directive 5000.01, *The Defense Acquisition System*, describes the management principles for DOD's acquisition programs. DOD Instruction 5000.02, *Operation of the Defense Acquisition System*, outlines a framework for managing acquisition programs. Collectively, these are known as the DOD 5000 series. Secretary of the Navy Instruction 5000.2E, *Department of the Navy Implementation and Operation of the Defense Acquisition System and the Joint Capabilities Integration and Development System*, issues procedures for major and non-major defense acquisition programs. Current law requires major defense acquisition programs to submit a summary report to several congressional committees at Milestone A that contains the total life-cycle cost estimate for the program. 10 U.S.C. § 2366a(c)(1)(B)(i).

However, the magnitude of CVN 65's anticipated cost of dismantlement and disposal is comparable to that of large DOD acquisition programs. Such programs generally are expected to provide more information to decision makers within DOD and Congress through formal reporting on plans, activities, and performance to support accountability than what has traditionally been provided with respect to Navy dismantlement and disposal activities.

The precedent-setting nature of the CVN 65 dismantlement and disposal adds a level of risk and heightens the importance of having sufficient accountability measures to facilitate oversight. There is greater potential for unexpected challenges to arise because a nuclear-powered aircraft carrier has not been dismantled and disposed of before. Additionally, CVN 65 provides an opportunity to establish a foundation for management and oversight of future aircraft carrier dismantlement and disposal efforts, with the first of 10 Nimitz-class carriers expected to reach the end of its service life in the next decade. Standards for internal control in federal government state that in order to identify and mitigate risk, program objectives such as a baseline for cost and schedule, should be clearly defined in measurable terms so performance in attempting to achieve those objectives can be assessed.[16] Doing so would also provide the Navy with the ability to collect important historical cost data that could be used to inform cost estimates for future aircraft carrier dismantlement and disposal efforts.

DOD acquisition programs could serve as a model to identify appropriate cost and schedule objectives for the CVN 65 dismantlement and disposal, even though it is not an acquisition program and not subject to these requirements. DOD acquisition programs with significant resource commitments comparable to that expected of CVN 65 are generally subject to structured oversight and have reporting requirements to support performance transparency and accountability.

---

[16] GAO-14-704G.

As discussed earlier, preliminary cost estimates for CVN 65 dismantlement and disposal may exceed $1 billion, regardless of the option the Navy ultimately selects. While many requirements for DOD acquisition programs are not relevant to dismantlement and disposal, even when costs may reach similar levels, we found elements of the reporting requirements associated with larger DOD acquisition category (ACAT) programs that the Navy could leverage to facilitate oversight of CVN 65 dismantlement and disposal.[17] For example, ACAT II programs—which have estimated costs comparable to CVN 65 dismantlement and disposal cost expectations—are required by DOD policy to establish a program cost and schedule baseline prior to program start and report any significant deviations from the established baseline.[18] They also are required by statute to provide information on risk management. Table 3 highlights some DOD acquisition program reporting elements that could support oversight of CVN 65 dismantlement and disposal.

For example, once a cost baseline is established, comparison to an independent cost estimate or assessment could provide greater assurance that the risks associated with performing CVN 65's large-scale, first-of-a-kind dismantlement activities were adequately considered and appropriately estimated. GAO's *Cost Estimating and Assessment Guide* states an independent review of a program's cost estimate is crucial to establishing confidence in the estimate.[19] It provides an unbiased test of whether the program cost estimate is reasonable and can be used to identify risks related to budget shortfalls or excesses.

---

[17] Defense acquisition program ACAT classifications depend on estimated costs and type of acquisition. All acquisition programs generally are required by statute or DOD guidance to provide program information at milestones and other decision points, although these requirements differ by ACAT level and may be tailored in many circumstances.

[18] Defense acquisition programs are designated as ACAT II when estimated to require a total expenditure for procurement of more than $835 million in fiscal year 2014 dollars.

[19] GAO, *GAO Cost Estimating and Assessment Guide: Best Practices for Developing and Managing Capital Program Costs*, GAO-09-3SP (Washington, D.C.: March 2009).

The Naval Center for Cost Analysis is responsible for developing independent cost assessments for ACAT II Navy programs, while the Office of the Secretary of Defense's Office of Cost Assessment and Program Evaluation develops independent cost estimates for major defense programs.

Table 3. DOD Acquisition Reporting Elements That Could Be Adapted for Use with CVN 65 Dismantlement and Disposal

Acquisition element	Description	Potential benefits of application
Acquisition program baseline	Establishes cost and schedule baseline estimates before project start and requires reporting of certain deviations in cost, schedule, or performance, among others	If a baseline for cost and schedule is established prior to contract award or funding naval shipyard work, it would allow tracking and reporting of cost and schedule deviations above certain thresholds from initial estimates through the life of the project
Independent cost estimate or assessment	Provides an independent evaluation of the quality, accuracy, and reasonableness of a program's cost estimate, with emphasis on specific cost and technical risks Identifies risks related to budget shortfalls or excesses	Allows for a validation of the Navy's cost expectations and a determination of whether the Navy's estimate accurately reflects the project expectations Could support a determination that the project can be completed without the need for significant adjustment to future budgets if the elements of project cost risk are properly identified, evaluated, and include mitigation plans
Risk management plan	Establishes procedures to identify, manage, and mitigate risks; estimates the likelihood and possible consequences in terms of cost, schedule, and performance	Procedures that could help to manage risks for a large-scale, first-of-its-kind project

Source: GAO analysis of DOD Instruction 5000.02 and GAO cost estimation guidance. | GAO-18-523.

As noted earlier, the Navy continues to refine its cost estimate of the naval shipyard option and expects to receive an estimate for the full commercial option from the Center for Naval Analyses in October 2018.

The Navy stated it considers the anticipated cost estimate and model from the Center for Naval Analyses to be the independent cost estimate for the full commercial option. We view this estimate as a valuable step in establishing cost expectations for the full commercial option, but believe it is inadequate because it will determine the Navy's cost expectations as opposed to validating an existing estimate—the intent of having an independent assessment. For the naval shipyard option, the Navy suggested no plans for an independent cost estimate as it continues to refine the current cost estimate prior to a decision for CVN 65.

Completing an independent cost estimate for both CVN 65 options prior to a Navy decision on its dismantlement and disposal approach would provide additional information to inform a decision that could have repercussions for carrier dismantlement and disposal activities for years to come. Adapting certain acquisition program requirements to the CVN 65 effort, as described above, would help the Navy establish baselines that can be tracked by decision makers to assess cost and schedule, and help identify deviations, if any. These types of reporting requirements would provide decision makers with greater information to support their oversight and hold the Navy accountable for meeting CVN 65 dismantlement and disposal expectations.

THE NAVY'S EVALUATION OF CVN 65 DISMANTLEMENT AND DISPOSAL OPTIONS IS HAMPERED BY A REGULATORY AUTHORITY DISAGREEMENT

The regulatory authority determines the rules, procedures, and oversight that will guide the dismantlement and disposal process for CVN 65. The Navy is considering three regulatory authority scenarios related to the naval shipyard or full commercial options, as discussed in table 4.

Table 4. Regulatory Authority Scenarios Being Considered by the Navy for Nuclear-Related CVN 65 Dismantlement and Disposal Activities

	Regulatory authority	Description of scenario
Naval shipyard option	Naval Reactors	Naval Reactors serves as regulatory authority for dismantlement and disposal managed by Puget Sound Naval Shipyard. Many existing shipyard oversight organizations and activities to assist Naval Reactors with its oversight based on extensive experience (about 130 reactors dismantled and disposed of at shipyard).
Full commercial option	Nuclear Regulatory Commission (NRC) or agreement state[a]	Naval Reactors officials view ship dismantlement and disposal as similar to its other nuclear-related processing and disposal activities, and believe NRC or a state should be able to serve as regulatory authority. NRC asserts it does not have regulatory authority for naval nuclear propulsion waste, nor would a state. Continued pursuit of the full commercial option without resolving this disagreement could present challenges for •ensuring accountability for safe dismantlement and disposal; •business case analysis used to examine costs, benefits, and risks; •Navy and industry cost estimates; and •industry workload and schedule estimates.
Full commercial option	Naval Reactors	Naval Reactors lacks experience, organizational structure, and practices to draw upon for nuclear-related dismantlement and disposal work performed by industry. Naval Reactors would need to establish new roles and responsibilities for oversight which it has typically done through naval shipyard support.

Source: GAO analysis of Naval Reactors and NRC information. | GAO-18-523

[a] NRC is authorized to enter into agreements where it relinquishes authority to states—referred to as agreement states—which then assume regulatory authority over certain nuclear materials.

The Navy Has Regulatory Precedent for the Naval Shipyard Option

If the Navy chooses the naval shipyard option, it can rely on Naval Reactors' extensive experience serving as the regulatory authority for dismantlement activities conducted at Puget Sound Naval Shipyard. Naval Reactors has overseen the dismantlement and disposal of roughly 130 reactors from submarines and cruisers by the naval shipyard. Many shipyard oversight organizations and activities, as well as on-site Naval Reactors personnel, help control environmental and human health exposures. For example, the Radiological Controls Office is responsible for monitoring radiation exposure to the workforce and ensuring radioactivity is confined to controlled work areas. The Nuclear Quality Division employs nuclear auditors who review performance, processes, and instructions for all nuclear work at the shipyard.

Although the scale and design of CVN 65 creates some unique dismantlement and disposal considerations as compared to the submarine and cruiser activities, Navy officials stated they plan to use the same organizations located at the shipyard and practices to oversee performance if they decide to complete the CVN 65 work at the shipyard. The environmental impact statement planned for CVN 65 is expected to outline the different needs that the aircraft carrier presents for the dismantlement process and disposal path, such as changes related to the transportation of CVN 65 reactor packages required if the Navy chooses to use four larger dual reactor compartment disposal packages instead of eight single packages to dispose of the carrier's reactors. While the Navy can rely on familiar regulatory practices to support the naval shipyard option, as discussed earlier, this option includes potential workload and schedule disadvantages.

Disagreement Persists about the Appropriate Regulatory Authority for the Full Commercial Option

Naval Reactors' position is that a commercial company could dismantle and dispose of CVN 65 under the regulatory authority of NRC or an agreement state. According to Naval Reactors officials, the full commercial option would represent a continuation of Naval Reactors' long history of nuclear-related activities with vendors licensed and regulated by NRC or agreement states. For example, Naval Reactors officials noted they commonly have used facilities licensed by NRC or agreement states for a range of manufacturing, processing, and disposal activities available for naval nuclear materials. Naval Reactor officials specifically assert that, as CVN 65 has already been defueled, such a facility should be able to process the byproduct material on the ship.[20] However, NRC stated its disagreement that it or an agreement state is able to serve as the regulatory authority for CVN 65, emphasizing that regulatory responsibility for the safe processing and disposal of Navy ships falls to Naval Reactors under its Department of Energy authority. NRC officials also noted that Naval Reactors has been regulating nuclear-powered ship dismantlement and disposal activities exclusively at Puget Sound Navy Shipyard for decades.

Coordination between Naval Reactors and NRC to identify the applicable regulatory authority and establish a regulatory plan for the CVN 65 full commercial option would help ensure accountability for safe dismantlement and disposal of CVN 65 under the full commercial option. It would also enable the Navy and commercial companies to effectively estimate costs. Without a resolution, the Navy could face challenges in estimating the cost and completing a comprehensive business case analysis of costs, benefits, and risks for the full commercial option if it is unsure of which regulatory authority will be responsible for enforcement. Furthermore, companies with potential interest in the CVN 65 work may not be able to effectively estimate the workload and associated cost without a clearly identified regulatory authority. Resolution of this

---

[20] Byproduct material includes certain material made radioactive by exposure to the operation of a nuclear reactor.

disagreement also has relevance for other future ship dismantlement and disposal activities, such as with the Surface Ship Support Barge in the near term and the Nimitz-class aircraft carriers in the long term.

Agreement State Program

The Atomic Energy Act gives the Nuclear Regulatory Commission (NRC) authority over domestic industrial, medical, and research uses of radioactive materials. The act also authorizes NRC to enter into agreements with states (called agreement states) so they assume, and NRC relinquishes, regulatory authority over specified radioactive materials. Specifically, NRC is authorized to enter into agreements to allow states to assume regulatory authority over source, byproduct, and special nuclear materials in quantities insufficient to form a critical mass. NRC must find a state program adequate to protect public health and safety and compatible with NRC's program for regulating such materials before entering into these agreements. The mechanism for the transfer of NRC's authority to a state is an agreement signed by the governor of the state and the chair of the Commission.

Map of Agreement States

Source: GAO (data); Map Resources (map). | GAO-18-523.

Navy Surface Ship Support Barge

The Surface Ship Support Barge is a dockside refueling facility constructed from a converted Navy tanker vessel used to disassemble spent nuclear fuel for shipment within a water pool. Naval Reactors noted this facility is now obsolete, with no further use planned, and the Navy is interested in dismantling and disposing of it commercially. According to Naval Reactors officials, the barge contains very low radioactivity in the water pool and fluid systems, which requires appropriate dismantlement and disposal measures.

The Navy halted its pursuit of a contract award to dismantle and dispose of the barge in early 2017 based on NRC formally stating it has no regulatory authority over the dismantlement and disposal of naval vessels. A Naval Reactors official stated a request for information may be issued in 2018 to solicit input from commercial companies for dismantlement and disposal of this barge, but plans remain unsettled.

Source: GAO analysis of Navy information. | GAO-18-523.

Naval Reactors' Position

Naval Reactors could use its own authority to regulate a full commercial dismantlement of CVN 65. Naval Reactors officials stated, however, that NRC or agreement states—which regulate industrial, medical, and research uses of radioactive materials—also have authority to regulate commercial dismantlement and disposal of CVN 65, and the Navy would benefit from leveraging their regulatory experience and structure. In particular, Naval Reactors officials stated that for the full commercial option, their responsibility to provide for processing and disposal of the byproduct material—which Naval Reactors indicated is what remains on CVN 65—can be best met by contracting with commercial companies licensed by NRC or an agreement state.[21]

---

[21] The Atomic Energy Act authorizes the Atomic Energy Commission to provide for safe processing and disposal of waste resulting from naval nuclear propulsion programs. When the Atomic Energy Commission was abolished, NRC received its licensing and related regulatory functions while all other authorities generally devolved to the Department of Energy. Because the Navy's use of CVN 65 is exempt from licensing by NRC, the Department of Energy bears the responsibility to provide for its processing and disposal. As

According to Naval Reactors officials, even if NRC maintains that it cannot regulate material from CVN 65, some states may do so under their own authority. Specifically, Naval Reactors' position is that states that had agreements with the old Atomic Energy Commission prior to its abolishment and the creation of NRC in 1974 were granted—and continue to maintain—authority to process naval nuclear propulsion waste. Accordingly, Naval Reactors officials stated that these states could serve as the sole regulatory authority over commercially-performed CVN 65 dismantlement and disposal.

Naval Reactors officials also asserted specific potential advantages of having NRC or an agreement state regulate commercial dismantlement and disposal of CVN 65. First, they said the regulatory structure that NRC and agreement states apply to commercial nuclear-related activities includes an enforcement process to impose fines for violations, which Naval Reactors does not have.[22]

Additionally, Naval Reactors officials noted the Navy's contract strategy options could be improved if NRC or an agreement state serves as the regulatory authority for CVN 65. Specifically, they stated that a reason for the Navy's interest in using NRC or agreement state authority is the possibility of emulating the firm-fixed-price contract currently being used to decommission a commercial nuclear power plant. In this example, the operating license was transferred from the utility that owns the plant and site to a dismantlement contractor to more quickly complete the decommissioning. This effectively gave a dismantlement and disposal company the power plant owner's responsibility for the safe dismantlement and disposal of the power plant, with NRC continuing to act as the regulatory authority. According to Naval Reactors, the firmfixed-price contract used in this case was viable because the dismantlement contractor had total responsibility independent of the plant owner to perform the work

---

a joint program of the Department of Energy and the Navy, Naval Reactors carries out this responsibility for the Department of Energy.

[22] In reviewing a draft of this report, DOD commented that Naval Reactors has the authority, deriving from the Department of Energy, to levy fines under the Price Anderson Amendments Act of 1988, but Naval Reactors has chosen not to utilize that authority. See 42 U.S.C. § 2273(a) (2018).

in accordance with the regulations and requirements of NRC. Naval Reactors officials stated that the firm-fixed-price contract created an incentive for the company to thoroughly understand what the work entailed and perform the work efficiently to maximize its profit.

Naval Reactors officials stated that a total separation of the owner from regulatory decisions and interpretations, like the one currently being used for the commercial nuclear power plant, is the Navy's best means to facilitate the potential use of a firm-fixed-price dismantlement contract for CVN 65. They further stated that an approach wherein Naval Reactors retained regulatory authority could undercut the prospect of a firm-fixed-price contract by eliminating the clear division between regulator and owner. In taking this position, Naval Reactors officials suggest that a conflict of interest exists in being both the owner who wants to establish a fixed price for the work as well as the regulator with the potential to affect costs. Naval Reactors officials also noted that if Naval Reactors were the regulator, with no experience in regulating this type of work, commercial companies could have difficulty pricing such regulatory risk. In contrast, they stated that in NRC-regulated commercial plant dismantlement, as well as agreement state-regulated, large-scale radioactive waste processing work, commercial companies have demonstrated that they are willing to accept this regulatory risk, agreeing to contracts on a fixed-price basis.

Nuclear Regulatory Commission's Position

In February 2017, NRC formally stated its position in a letter responding to a congressional inquiry that it has no regulatory authority for Navy ships, such as CVN 65. NRC said that under the Atomic Energy Act it is the responsibility of the Department of Energy, and accordingly Naval Reactors, to provide for processing and disposal of naval nuclear propulsion waste.[23] NRC stated that agreement states also lack jurisdiction because their authority derives from NRC's authority. NRC officials we interviewed also disputed Naval Reactors' position that states have independent authority to process naval nuclear propulsion waste for two

---

[23] As noted previously, this Department of Energy Authority has been delegated to Naval Reactors by Executive Order 12344.

reasons. First, they stated that regulation of reactor dismantlement is not an activity that can be relinquished to the states.[24] Second, they pointed out that the authorities that can be relinquished to the states under the Atomic Energy Act are licensing activities conducted under specific provisions of the act, and that the responsibility to safely process and dispose of naval nuclear propulsion waste is conducted under a different set of provisions which are not subject to licensing.[25]

NRC officials acknowledged that naval nuclear propulsion waste has been processed at facilities licensed by NRC or an agreement state, but distinguished such examples from CVN 65. Specifically, they noted that no additional regulatory oversight was required to process incidental amounts of such waste at facilities licensed to process commercial waste, but CVN 65 is not licensed by NRC or an agreement state and would involve only naval nuclear propulsion waste. NRC officials emphasized that the additional work that would be required to regulate the dismantlement of CVN 65—an unlicensed facility—puts it beyond NRC's jurisdiction. Additionally, NRC stated that while such work could be carried out by a contractor, including a contractor with an NRC or agreement state license, the work would not be covered by that license, as NRC and agreement states do not have authority to regulate such activity. Essentially, NRC's position is that while Naval Reactors can contract to have the dismantlement and disposal performed by a commercial entity, Naval

---

[24] The Atomic Energy Act prohibits relinquishment to states of authorities associated with the construction and operation of nuclear reactors, and NRC's position is that dismantlement would constitute construction and/or operation. Naval Reactors contends that CVN 65's reactors have been rendered inoperable by the defueling and deactivation process to the extent that they no longer constitute reactors.

[25] The Atomic Energy Act allows for the relinquishment to states of authorities in chapters 6, 7, and 8 of the act related to licensing byproduct materials, source materials, and special nuclear materials in quantities not sufficient to form a critical mass. Naval Reactors' responsibility to provide for the safe processing and disposal of naval propulsion waste is located in chapter 9 of the act and is not subject to the licensing requirements of the act. The act also allows for the relinquishment of authorities under section 161 of the act, and section 161(b) of the act authorizes the establishment of standards governing the possession of byproduct material without regard to licensing. However, neither Naval Reactors nor NRC had analyzed the extent to which this provision was previously used by the Atomic Energy Commission or could be used by Naval Reactors to delegate authority to states over naval nuclear propulsion waste.

Reactors would retain its own regulatory responsibility for enforcing that contract.

NRC stated that if Naval Reactors desired technical support in regulating a commercial dismantlement, NRC or an agreement state could provide such services through a contract. This approach, according to NRC officials, would offer Naval Reactors a regulatory consultant familiar with commercial dismantlement while maintaining Naval Reactors as the regulatory enforcement authority. In such an arrangement, NRC or an agreement state could identify regulatory concerns, but Naval Reactors would be responsible for determining what corrective action is taken to address those concerns. Naval Reactors officials stated they are in ongoing discussions with NRC about this potential approach. They also asserted that this potential approach is not optimal because, as previously discussed, it could create regulatory uncertainty for commercial companies by preventing a clear separation of the regulator and owner.

Naval Reactors Lacks Regulatory Experience and Structure for Commercial Dismantlement and Disposal

Since Naval Reactors has its own authority as part of the Department of Energy, it could choose to regulate a CVN 65 commercial dismantlement. However, with Puget Sound Naval Shipyard having performed the dismantlement and disposal work for previous nuclear-powered vessels, Naval Reactors lacks experience to draw upon for a full commercial option. It also cannot rely on the organizational structure and practices in place at Puget Sound Naval Shipyard to support a commercial CVN 65 dismantlement that will be conducted at an offsite facility. If Naval Reactors serves as the regulatory authority for a full commercial dismantlement, it will have to determine what mechanisms are needed to provide sufficient monitoring of the work and how they will fulfill the roles and responsibilities typically filled by the naval shipyard's support. These mechanisms may include elements similar to those used by the naval shipyard as well as new ones unique to the dismantlement practices used by commercial companies.

A significant consideration for Naval Reactors when working to establish an approach to monitor commercial dismantlement and disposal is the component-based dismantlement process that companies may use. This process, which is commonly used to dismantle commercial nuclear power reactors, involves segmenting reactor components (i.e., cutting to reduce in size) so the pieces can be put in standardized containers for transport and disposal. This process is a contrast to the traditional dismantlement approach that Nuclear Reactors uses at Puget Sound Naval Shipyard— an approach that would leave CVN 65's reactors largely intact by encasing them in packages for disposal. As noted by Naval Reactors officials, commercial dismantlement practices potentially could require the Navy to decide whether to adjust its standard radiological work practices to better align with different dismantlement and disposal activities or use the same practices it uses for work performed at Puget Sound Naval Shipyard. Using the same practices could affect cost expectations for the Navy and commercial companies by changing the way the work is performed. As an example, applying the Navy's standard practices for total containment of radionuclides to a dismantlement process that involves increased cutting could require additional measures to control the work environment.

CONCLUSION

Over 50 years ago, CVN 65 set a precedent as the Navy's first nuclear-powered aircraft carrier. The Navy's plans and decisions for this aircraft carrier's dismantlement and disposal represent an opportunity to create a standard that the Navy may use for decades to come as the Nimitz-class carriers enter retirement. As the Navy considers how to proceed, it will be critical to ensure that there is sufficient oversight and accountability for what likely will be an effort greater than $1 billion that lasts the better part of a decade. Since budget exhibits are a primary tool to aid Congress in making well-informed funding decisions, without additional details, transparency and the ability to assess CVN 65 progress could be limited. In particular, a more robust budget exhibit for CVN 65 that includes cost and schedule information across the Future Years Defense Program, as well as the status of activities—including any contract awards and a tracking of high level changes in cost and schedule—could help increase transparency for oversight.

Reporting requirements for DOD acquisition programs, which are not required or currently planned for CVN 65 dismantlement and disposal, provide examples of the types of information that decision makers can use to ensure that resource-intensive programs are meeting expectations or make changes as necessary. Without establishing a cost and schedule baseline that has been validated by an independent cost estimate or assessment, it will be difficult for decision makers to track cost and schedule performance or have confidence in CVN 65 costs. The Navy has indicated it is receptive to providing additional information to support oversight that is commensurate with other Navy programs of a similar funding level. However, the Navy also stated that it requires clear direction from DOD leadership or Congress on what additional accountability measures are desired before it would make any changes to current budget exhibits and reporting.

Naval Reactors is charged with cradle-to-grave responsibility for our nation's naval nuclear propulsion material. The disagreement between Naval Reactors and NRC about the regulatory authority for commercial dismantlement and disposal of Navy nuclear ships persists. Coordination between the two agencies to identify the applicable regulatory authority for a full commercial dismantlement and disposal of CVN 65 and to develop a regulatory plan would help establish which practices and standards will apply to uphold nuclear safety and security. It would also help ensure the Navy's selection of a dismantlement and disposal plan for CVN 65 is informed by well understood regulatory expectations and cost and schedule estimates that reflect those expectations.

Matter for Congressional Consideration

We are making one matter for congressional consideration.

Congress should consider requiring Naval Reactors to coordinate with the Nuclear Regulatory Commission for any CVN 65 dismantlement and disposal performed commercially to identify the applicable regulatory authority. In the event that an entity other than Naval Reactors will serve as the regulatory authority, Naval Reactors should submit to Congress a plan that identifies the regulatory authority for CVN 65 activities, and includes acknowledgement from that regulatory entity of its agreement with Naval Reactors and the legal basis for its authority. If the regulatory entity is an agreement state, such acknowledgment should be coordinated with the Nuclear Regulatory Commission. (Matter 1)

Recommendations for Executive Action

We are making the following four recommendations to DOD.

The Secretary of Defense should ensure that the Navy provides additional information in the annual President's budget exhibits associated with CVN 65 dismantlement and disposal to facilitate improved transparency and accountability. Additions should, at a minimum, include the CVN 65 funding estimate across the Future Years Defense Program, activities planned or performed for applicable fiscal years, tracking of total cost and high level changes in cost and schedule from the prior year with explanations for changes, and if applicable, contract type, awardee, award value, and award and completion date estimates. (Recommendation 1)

The Secretary of Defense should require the Navy to obtain an independent cost estimate, performed by DOD's Office of Cost Analysis and Program Evaluation or the Naval Center for Cost Analysis, for both the naval shipyard and full commercial options before choosing a dismantlement and disposal approach for CVN 65. (Recommendation 2)

The Secretary of Defense should require the Navy to complete a risk management plan prior to beginning the CVN 65 dismantlement and disposal. (Recommendation 3)

The Secretary of Defense should require the Navy to approve a cost and schedule baseline prior to beginning the CVN 65 dismantlement and disposal. (Recommendation 4)

AGENCY COMMENTS

We provided a draft of this chapter to DOD and NRC for comment. Both DOD and NRC agreed with the draft report and its findings, and DOD concurred with the four recommendations we directed to the department. DOD and NRC provided written comments, which have been reproduced in appendix II and appendix III, respectively. DOD and NRC also provided technical comments, which we incorporated as appropriate.

Shelby S. Oakley
Director, Contracting and National Security Acquisitions

List of Committees

The Honorable John McCain
Chairman

The Honorable Jack Reed
Ranking Member
Committee on Armed Services
United States Senate

The Honorable Richard Shelby
Chairman

The Honorable Dick Durbin
Ranking Member
Subcommittee on Defense Committee on Appropriations
United States Senate

The Honorable Mac Thornberry
Chairman

The Honorable Adam Smith
Ranking Member
Committee on Armed Services
House of Representatives

The Honorable Kay Granger
Chairwoman

The Honorable Pete Visclosky
Ranking Member Subcommittee on
Defense Committee on Appropriations
House of Representatives

APPENDIX I – OBJECTIVES, SCOPE, AND METHODOLOGY

This chapter (1) describes the differences between the dismantlement and disposal options under consideration, including cost and schedule as well as workload and facilities; (2) evaluates the Navy's funding and reporting practices for dismantlement and disposal activities; and (3) assesses the effect that nuclear regulatory authority considerations have on dismantlement and disposal options for CVN 65.

To identify the differences between the potential CVN 65 dismantlement and disposal options, we reviewed Navy documentation on prior, ongoing, and future dismantlement and disposal activities, as well as information related to the different options the Navy has considered or is considering for CVN 65. We interviewed Navy officials and reviewed documentation from the Naval Sea Systems Command, which includes Naval Reactors, and Puget Sound Naval Shipyard and Intermediate Maintenance Facility (hereafter referred to as Puget Sound Naval Shipyard). To obtain an understanding of the full commercial dismantlement and disposal approach, including work practices and potential work sites, we interviewed officials and reviewed documentation from commercial companies that the Navy identified as having involvement in shipbreaking or nuclear-related industries and potential interest in CVN 65. These companies include Atkins Global; Energy Solutions; Huntington Ingalls Industries (HII – Nuclear); International Shipbreaking Limited; North Star Group Services; and Waste Control Specialists.

For CVN 65 cost and schedule estimates, the Navy considers all estimates to still be preliminary because the Navy has yet to formally begin the environmental impact statement process and remains years away from a decision on its dismantlement and disposal approach. As a result, we did not formally evaluate the reasonableness of any cost or schedule estimates. However, we did review the initial estimates to gain insight on historical and current cost expectations. To assess the Navy's preliminary cost estimates for the naval shipyard option, we reviewed Navy data on the basis for the cost estimates, particularly estimates since 2011. This included reviewing the cost factors that contributed to each estimate to understand how the shipyard's increasing knowledge of CVN 65's ship characteristics and changes to the planned dismantlement approach fed into the different estimates. For the Navy's notional cost estimate of the CVN 65 full commercial option, we reviewed the data and approach used by the Navy to develop initial cost information. This included commercial decommissioning data, which the Navy used to establish a rough order of magnitude cost estimate based on the limited information available that is comparable to CVN 65 dismantlement and disposal.

We used the same data to generate our own notional estimated cost range based on a Nuclear Regulatory Commission (NRC) cost formula, as well as published data from the Organisation for Economic Cooperation and Development's Nuclear Energy Agency. This included analysis of costs reported by operating power reactor licensees in NRC's 2015 decommissioning funding status report to comply with decommissioning financial assurance reporting requirements. Our review of historical data from the Nuclear Energy Agency and a 2011 report on nuclear decommissioning by an independent panel established by the California Public Utilities Commission helped us identify cost drivers and categories of costs attributed to specific activities that occur when decommissioning commercial power plants.

To assess workload and facility considerations related to CVN 65, we analyzed Puget Sound Naval Shipyard workload and resource requirements data for fiscal years 2006 through 2025, and facility data for fiscal years 2018 through 2035. To assess the reliability of these data, we interviewed knowledgeable officials and reviewed documentation to verify the controls and measures used to validate and maintain the data. We determined these data to be reliable for our purposes of discussing the existing and planned workload at Puget Sound Naval Shipyard. We compared projections to actual workload when available to identify differences and compared the average amount of annual projected workload to the average amount of annual projected workforce available. We also reviewed a 2018 report on the Navy's strategic plan for addressing the infrastructure deficiencies at the public naval shipyards as well as the Navy's long-range shipbuilding plans for fiscal years 2011, 2016, and 2019. Additionally, we reviewed past GAO reports that addressed operation and maintenance activities at naval shipyards, and the related workload demands and facilities' requirements.[26]

[26] GAO, *Naval Shipyards: Actions Needed to Improve Poor Conditions that Affect Operations*, GAO-17-548 (Washington, D.C.: Sept. 12, 2017) and *Depot Maintenance: Executed*

To identify the Navy's funding and reporting practices for dismantlement and disposal activities, we reviewed Navy documentation on prior, ongoing, and future ship dismantlement and disposal activities, as well as Navy procurement and operation and maintenance budget exhibits— fiscal years 2016 and 2017 for procurement exhibits and fiscal years 2007 through 2017 for operation and maintenance budget exhibits. We also reviewed Federal Acquisition Regulations, Office of Management and Budget guidance on budget information, and the Department of Defense and Navy acquisition regulations. We interviewed officials from Naval Reactors and the Program Executive Office for Aircraft Carriers. Based on these efforts, we evaluated the Navy's historical approach for funding, conducting oversight, and reporting on dismantlement and disposal activities. We assessed the Navy's approach against federal standards for internal control.[27] Additionally, we assessed how funding and typical reporting requirements for Department of Defense acquisition programs align with the potential need to facilitate oversight for CVN 65 dismantlement and disposal.

To determine the effect that nuclear regulatory authority considerations have on dismantlement and disposal for CVN 65, we examined applicable laws, regulations, executive orders, policies, and guidance documents related to nuclear-powered ships. We also reviewed past GAO reports related to environmental and nuclear requirements.[28] We reviewed Navy documentation on prior, ongoing, and future ship dismantlement and disposal activities. We also interviewed officials and reviewed documentation from Naval Reactors; the Assistant Secretary of the Navy for Energy, Installations, and Environment; the Chief of Naval Operations Environmental Readiness Division; the Program Executive Office for

---

Workload and Maintenance Operations at DOD Depots, GAO-17-82R (Washington, D.C.: Feb. 3, 2017).

[27] GAO, *Standards for Internal Control in the Federal Government*, GAO-14-704G (Washington, D.C.: September 2014).

[28] GAO, *Nuclear Regulatory Commission: Regulatory Fee-Setting Calculations Need Greater Transparency*, GAO-17-232 (Washington, D.C.: Feb. 2, 2017); *Nuclear Security: NRC Has Enhanced the Controls of Dangerous Radioactive Materials, but Vulnerabilities Remain*, GAO-16-330 (Washington, D.C.: July 1, 2016); and *National Environmental Policy Act: Little Information Exists on NEPA Analyses*, GAO-14-370 (Washington, D.C.: Apr. 15, 2014).

Aircraft Carriers; Puget Sound Naval Shipyard; and the Nuclear Regulatory Commission. Additionally, we interviewed officials from the Washington State Departments of Health and Ecology, the Texas Commission on Environmental Quality, and the Texas Department of State Health Services—two states in which ship dismantlement activities have recently occurred and that have nuclear waste disposal sites.

We conducted this performance audit from August 2017 to August 2018 in accordance with generally accepted government auditing standards. Those standards require that we plan and perform the audit to obtain sufficient, appropriate evidence to provide a reasonable basis for our findings and conclusions based on our audit objectives. We believe that the evidence obtained provides a reasonable basis for our findings and conclusions based on our audit objectives.

APPENDIX II – COMMENTS FROM THE DEPARTMENT OF DEFENSE

THE ASSISTANT SECRETARY OF DEFENSE
3015 DEFENSE PENTAGON
WASHINGTON, DC 20301-3015

ACQUISITION

JUL 19 2018

Ms. Shelby Oakley
Director, Contracting and National Security
U.S. Government Accountability Office
441 G Street, NW
Washington, DC 20548

Dear Ms. Oakley:

This is the Department of Defense (DoD) response to the GAO Draft Report, GAO-18-523, 'AIRCRAFT CARRIER DISMANTLEMENT AND DISPOSAL: Options Warrant Additional Oversight and Raise Regulatory Questions,' dated June 18, 2018 (GAO Code 102258). Detailed responses to the report recommendations are enclosed.

In addition to the four recommendations, the GAO Draft Report recommended that Congress consider requiring Naval Reactors to coordinate with the Nuclear Regulatory Commission (NRC) for any CVN 65 dismantlement and disposal performed commercially to identify the applicable regulatory authority. Naval Reactors has been coordinating with the NRC on this matter for nearly two years. NRC believes it does not possess regulatory authority to govern dismantlement of CVN 65 by commercial entities. If this is accurate, the Navy would not be able to contract on a commercial basis for radiological dismantlement services for CVN 65 that are regulated by the NRC and NRC Agreement States. The Department is concerned that such an outcome effectively deprives the Navy of access to the same commercially available, safe, and well NRC-regulated services which the private sector may use, potentially affecting the Department's ability to efficiently balance industrial workload and make maximum use of the funds provided by Congress for the National Defense by requiring a less efficient and duplicative regulatory framework. Since the concern expressed by NRC is a lack of authority, and such a position severly impacts the Navy's ability to pursue a commercial option, we are highlighting it in our reponse while continuing to work with NRC to resolve.

Naval Reactors will continue to coordinate with the NRC and is currently engaged in developing a novel regulatory oversight arrangement concerning the Surface Ship Support Barge, a smaller radiological dismantlement project, which, if adopted, may serve as guidance for potential use in larger disposal applications.

Finally, I would like to commend the GAO on the thoroughness of this review. This is a complex topic that has not been previously addressed. The GAO's investigators have been diligent in seeking all of the facts and perspectives from a wide range of people and organizations. Within the DoD, we have had very good discussions with GAO personnel, and we are satisfied that our concerns and opinions have been fairly considered. Hopefully, this GAO review will be helpful in moving forward on a matter that may have a major impact on the cost and efficiency of DoD radiological disposal.

Sincerely,

Kevin M. Fahey

GAO DRAFT REPORT DATED JUNE 18, 2018
GAO-18-523 (GAO CODE 102258)

"AIRCRAFT CARRIER DISMANTLEMENT AND DISPOSAL: OPTIONS WARRANT ADDITIONAL OVERSIGHT AND RAISE REGULATORY QUESTIONS"

DEPARTMENT OF DEFENSE COMMENTS TO THE GAO RECOMMENDATIONS

RECOMMENDATION 1: The Secretary of Defense should ensure that the Navy provides additional information in the annual President's budget exhibits associated with CVN 65 dismantlement and disposal to facilitate improved transparency and accountability. Additions should, at a minimum, include the CVN 65 funding estimate across the Future Years Defense Program, activities planned or performed for applicable fiscal years, tracking of total cost and high level changes in cost and schedule from the prior year with explanations for changes, and if applicable, contract type, awardee, award value, and award and completion date estimates.

DoD RESPONSE: Concur. The Department agrees that the current format of budget exhibits associated with CVN 65 dismantlement and disposal limits transparency and accountability for the level of investment required for the effort. In accordance with this letter, the Navy agrees to develop and submit additional cost, schedule, and contract information to supplement existing budget exhibits and to continue such reporting through completion of the dismantlement and disposal of CVN 65. The information to be provided will include:

- A summary of activities and significant developments in connection with the dismantlement and disposal.
- A detailed description of cost and schedule performance against the baselines established for the dismantlement and disposal, including a description of and explanation for any variance from the cost or schedule baseline.

- A description of the amounts requested, or intended or estimated to be requested, for the dismantlement and disposal for each fiscal year covered by the future years defense program and any fiscal years prior to or after the period of the future-years defense program.

For reference purposes, this item will be identified as GAO-18-523-01.

RECOMMENDATION 2: The Secretary of Defense should require the Navy to obtain an independent cost estimate, performed by DoD's Office of Cost Analysis and Program Evaluation or the Naval Center for Cost Analysis, for both the naval shipyard and full commercial options before choosing a dismantlement and disposal approach for CVN 65.

DoD RESPONSE: Concur. The Department agrees that the Navy should obtain an independent cost estimate for both the naval shipyard and the full commercial dismantlement and disposal options. In accordance with this letter, the Navy agrees to obtain both independent cost estimates through the Office of Cost Analysis and Program Evaluation (CAPE). The CAPE independent cost estimates will be completed in time to support the Navy's final decision on which option, naval shipyard or full commercial, will be pursued. For reference purposes, this item will be identified as GAO-18-523-02.

RECOMMENDATION 3: The Secretary of Defense should require the Navy to complete a risk management plan prior to beginning CVN 65 dismantlement and disposal.

DoD RESPONSE: Concur. The Department agrees that a risk management plan is needed. In accordance with this letter, the Navy agrees to prepare a risk management plan covering the selected dismantlement and disposal option. The risk management plan will be completed not less than 90 days before the award of a contract to a commercial entity or the provision of funds to a government owned naval shipyard for the dismantlement and disposal of CVN 65. For reference purposes, this item will be identified as GAO-18-523-03.

RECOMMENDATION 4: The Secretary of Defense should require the Navy to approve a cost and schedule baseline prior to beginning the CVN 65 dismantlement and disposal.

DoD RESPONSE: Concur. The Department agrees that a cost and schedule baseline is needed to ensure accountability in the CVN 65 dismantlement and disposal effort. In accordance with this letter, the Navy agrees to prepare and approve a cost and schedule baseline covering the selected dismantlement and disposal option and to monitor cost and schedule against the baseline for the duration of the dismantlement and disposal effort. The cost and schedule baseline will be approved not less than 90 days before the award of a contract to a commercial entity or the provision of funds to a government owned naval shipyard for the dismantlement and disposal of CVN 65. For reference purposes, this item will be identified as GAO-18-523-04.

APPENDIX III – COMMENTS FROM THE NUCLEAR REGULATORY COMMISSION

UNITED STATES
NUCLEAR REGULATORY COMMISSION
WASHINGTON, D.C. 20555-0001

July 18, 2018

Shelby S. Oakley, Director
Contracting and National Security Acquisitions
U.S. Government Accountability Office
441 G Street, NW
Washington, DC 20226

Dear Ms. Oakley:

On behalf of the U.S. Nuclear Regulatory Commission (NRC), I am responding to your email dated June 18, 2018, which provided the NRC with an opportunity to review and comment on the U.S. Government Accountability Office (GAO) draft report GAO-18-523, "Aircraft Carrier Dismantlement and Disposal: Options Warrant Additional Oversight and Raise Regulatory Questions."

The NRC appreciates the opportunity to review the draft report as well as the GAO staff's professionalism and many constructive interactions during this GAO engagement. Overall, the NRC agrees with the draft report and its findings. The draft report accurately describes the dismantlement and disposal options under consideration, the NRC's regulatory authority, and funding and reporting practices. In the enclosure to this letter, we have provided some minor comments and clarifications for your consideration.

Sincerely,

Margaret M. Doane

Margaret M. Doane
Executive Director
for Operations

In: Top 5 Reports and Testimonies … ISBN: 978-1-53615-511-2
Editors: C. Johnston and W. Payne

Chapter 3

CRIMINAL ALIEN STATISTICS: INFORMATION ON INCARCERATIONS, ARRESTS, CONVICTIONS, COSTS, AND REMOVALS*

United States Government Accountability Office

ABBREVIATIONS

ACRIMe	Alien Criminal Response Information Management System
BJA	Bureau of Justice Assistance
BOP	Bureau of Prisons
DHS	Department of Homeland Security
DOJ	Department of Justice
FBI	Federal Bureau of Investigation

* This is an edited, reformatted and augmented version of United States Government Accountability Office; Report to Congressional Requesters, Publication No. GAO-18-433, dated July 2018.

IAR	Immigration Alien Response
ICE	Immigration and Customs Enforcement
NSD	National Security Division
NGI	Next Generation Identification
PCEPI	Personal Consumption Expenditures Price Index
SCAAP	State Criminal Alien Assistance Program
USCIS	U.S. Citizenship and Immigration Services

WHY GAO DID THIS STUDY

As of 2014, DHS estimated the total alien—a person who is not a citizen or national of the United States— population in the United States was about 27.1 million. Members of the alien population that have been arrested and convicted of crimes in the United States are referred to as criminal aliens. The costs associated with incarcerating criminal aliens are borne by the federal government, states, and localities. GAO was asked to update its March 2011 report on criminal alien statistics.

This chapter addresses, among other things, the (1) number and nationality of incarcerated criminal aliens, (2) number of criminal alien arrests and convictions, (3) estimated costs associated with incarcerating criminal aliens, and (4) experiences of criminal aliens after incarceration in federal prison.

GAO analyzed the most recent data available on criminal aliens, generally from fiscal years 2010 through 2016. Specifically, GAO analyzed data for criminal aliens incarcerated in federal prisons from fiscal years 2011 through 2016, and this group is the federal criminal alien population studied for this chapter. GAO also analyzed data for certain criminal aliens incarcerated in state prisons and local jails that received SCAAP funding from fiscal years 2010 through 2015, and this group is referred to as the state and local study population reviewed for this chapter. SCAAP data represents a portion of all criminal aliens incarcerated in state prisons and local jails. GAO used SCAAP data because there are no reliable data available on all criminal aliens incarcerated in every U.S. state prison and local jail.

GAO analyzed (a) DOJ data on criminal alien incarcerations, arrests, convictions, and costs; (b) conviction data from the five state prison systems with the largest number of SCAAP criminal alien incarcerations in fiscal year 2015; and (c) DHS data on removals.

For its arrest analyses, GAO selected generalizable random samples of (1) 500 criminal aliens from about 197,000 that were incarcerated in federal prisons from fiscal years 2011 through 2016 and (2) 500 SCAAP criminal aliens from about 533,000 that were incarcerated in state prisons and local jails from fiscal years 2010 through 2015. These samples included only those criminal aliens who had a Federal Bureau of Investigation (FBI) number—a unique identifier. This unique identifier allowed GAO to obtain arrest/transfer histories for these criminal aliens from FBI's database, which includes data from law enforcement agencies across the nation.

While the samples selected for the arrest analyses allowed GAO to estimate and provide insights about the arrest history of the criminal aliens in the study populations, these findings are not generalizable to the arrest history of criminal aliens not included in these populations. These data did not allow GAO to distinguish between a new arrest and a transfer from one agency to another; therefore, these are collectively referred to as "arrests/transfers." An arrest/transfer can be for multiple offenses. GAO's arrest analyses have a margin of error of plus or minus 7 percentage points or fewer. For GAO's analyses of state conviction data, information obtained from the selected state prison systems is not generalizable to all state prison systems, but provides useful insights on why criminal aliens were incarcerated.

What GAO Found

From fiscal years 2011 through 2016, the criminal alien proportion of the total estimated federal inmate population generally decreased, from about 25 percent to 21 percent (as shown in the figure below). During this period, the estimated number of criminal aliens incarcerated in federal prisons decreased from about 50,400 to about 39,500, or 22 percent. Ninety-one percent of these criminal aliens were citizens of one of six countries, including Mexico, Honduras, El Salvador, Dominican Republic, Colombia, and Guatemala.

Based on data from the Department of Justice's (DOJ) State Criminal Alien Assistance Program (SCAAP), which reimburses states and localities for a portion of criminal alien incarceration costs, the number of SCAAP criminal alien incarcerations in state prisons and local jails that received SCAAP reimbursements also decreased from about 282,300 in fiscal year 2010 to about 169,300 in fiscal year 2015, or 40 percent. The decrease can be attributed to (1) general declines in the number of SCAAP criminal alien incarcerations in each of the participating state prisons and local jails that participated in SCAAP, and (2) a reduction in the number of states and localities that participated in SCAAP. Seventy-six percent of SCAAP criminal alien incarcerations in fiscal year 2015 were born in one of six countries, including Mexico, El Salvador, Honduras, Guatemala, Cuba, and Germany.

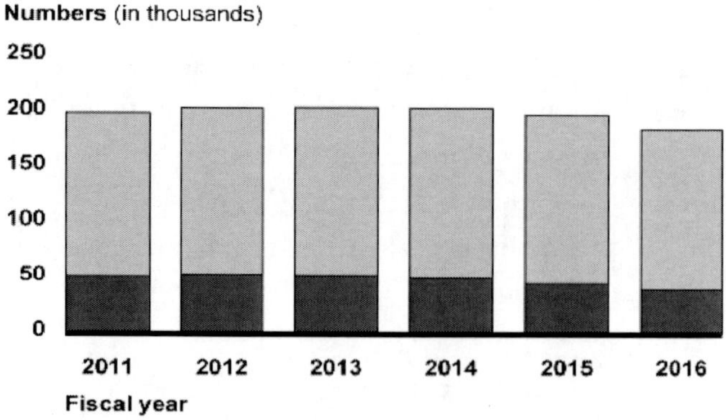

Source: GAO analysis of Bureau of Prisons data. | GAO-18-433.

Estimated Number of Individuals, by Citizenship, Incarcerated in Federal Prisons from Fiscal Years 2011 through 2016.

Based on a random sample of criminal aliens incarcerated in federal prisons during fiscal years 2011 through 2016 and based on a random sample of SCAAP criminal aliens incarcerated in state prisons and local jails during fiscal years 2010 through 2015, GAO estimated the following:

- The approximately 197,000 federal criminal aliens included in GAO's analysis were arrested/transferred about 1.4 million times for approximately 2 million offenses from over 43 years (from 1974 through 2017); 42 percent of the offenses that these criminal aliens were arrested for were related to immigration and 26 percent were related to drugs or traffic violations.
- The approximately 533,000 SCAAP criminal aliens included in GAO's analysis were arrested/transferred about 3.5 million times for approximately 5.5 million offenses from over 53 years (from 1964 through 2017); 52 percent of the offenses that these SCAAP criminal aliens were arrested for were related to traffic violations, drug offenses, or immigration offenses.

An arrest does not necessarily result in prosecution or a conviction of all, or any, of the offenses for which an individual is arrested. GAO's analyses found that 92 percent of the criminal aliens incarcerated in federal prison from fiscal years 2011 through 2016 were convicted of primary offenses related to immigration or drugs—a primary offense is the one with the longest maximum sentence, as determined by the relevant agency. At the state level, SCAAP criminal aliens incarcerated in fiscal year 2015 in Arizona, California, Florida, New York, and Texas state prison systems were convicted of various primary offenses. While the most common primary offenses varied by each of the five states, they generally related to drug, homicide, or sex offenses.

GAO's analyses found that the total annual estimated federal costs to incarcerate criminal aliens decreased from about $1.56 billion to about $1.42 billion from fiscal years 2010 through 2015. These costs included federal prison costs and reimbursements to state prison and local jail systems for a portion of their costs. GAO's analyses also show that selected annual estimated operating costs of state prison systems to incarcerate SCAAP criminal aliens decreased from about $1.17 billion to about $1.11 billion from fiscal years 2010 through 2015. These selected costs included correctional officer salaries, medical care, food service, and utilities.

Of the approximately 165,700 criminal aliens who completed a term of incarceration in federal prison from fiscal years 2011 through 2016, about 157,400 or 95 percent were subsequently removed from the United States by the Department of Homeland Security (DHS). The majority (about 146,500) of the criminal aliens who completed a term of federal prison incarceration did not have a subsequent reincarceration in a federal prison. However, about 19,300 were subsequently reincarcerated in a federal prison at least once and about 5,500 were reincarcerated in a state prison or local jail system that received SCAAP funding. These experiences after federal prison incarceration are not mutually exclusive. For example, criminal aliens could have been removed from the United States by DHS after their incarceration in federal prison, then reentered the United States and subsequently become reincarcerated in either a federal or state prison or local jail.

Federal Prison Reincarcerations of Criminal Aliens, Fiscal Years 2011 through 2016

Number of federal prison reincarcerations	Number of criminal aliens	Percent
0	146,500	88.4
1	16,700	10.1
2	2,200	1.3
3	300	0.2
4 or more	<100	<0.1
Total[a]	165,700	100

Source: GAO analysis of Bureau of Prisons data. | GAO-18-433.

[a]Numbers may not sum to totals due to rounding. Percentages may not add to 100 due to rounding.

July 17, 2018

The Honorable Charles E. Grassley
Chairman
Committee on the Judiciary
United States Senate

The Honorable Steve King
House of Representatives

The Honorable Pete Sessions
House of Representatives

In 2014, the Department of Homeland Security (DHS) estimated the total alien population in the United States was about 27.1 million. Of that number, DHS reported that about 12.1 million aliens were without lawful status or presence.[1] The Immigration and Nationality Act defines an alien as a person who is not a citizen or national of the United States.[2] Aliens, with or without lawful status, who have been arrested and convicted of crimes are known as criminal aliens.[3] U.S. Immigration and Customs

---

[1] According to DHS, the remaining approximately 15 million aliens includes lawful permanent residents (13.2 million), resident nonimmigrants (1.7 million), and individuals granted refugee or asylee status (0.1 million), as of 2014. Data on alien populations come from DHS's Office of Immigration Statistics, see DHS Office of Immigration Statistics, *Estimates of the Lawful Permanent Resident Population in the United States: January 2014* (Washington, D.C.: June 2017); *Estimates of the Size and Characteristics of the Resident Nonimmigrant Population in the United States: Fiscal Year 2014* (Washington, D.C.: December 2016); *Refugees and Asylees: 2014* (Washington, D.C.: April 2016); and *Estimates of the Unauthorized Immigrant Population Residing in the United States: January 2014* (Washington, D.C.: July 2017). DHS's *Estimates of the Unauthorized Immigrant Population Residing in the United States: January 2014* is the most recent report that DHS issued on this population.

[2] See 8 U.S.C. § 1101(a)(3), (a)(22). A "national of the United States" means a U.S. citizen, or a person who, though not a U.S. citizen, owes permanent allegiance to the United States, which can include individuals who were born in American Samoa or in the Commonwealth of the Northern Mariana Islands who have chosen to be U.S. nationals instead of U.S. citizens.

[3] For our report purposes, criminal aliens are aliens with or without lawful status convicted in the United States of crimes. According to DOJ officials, other entities may define criminal

Enforcement (ICE), one of DHS's component agencies, is responsible for identifying, apprehending, detaining, litigating charges of removability against, and removing aliens, including criminal aliens, who are in the United States in violation of U.S. immigration law.[4] The costs associated with incarcerating criminal aliens are borne by the federal government as well as states and localities.[5] Aliens convicted in federal court and sentenced to a term of imprisonment are committed to the custody of the Department of Justice's (DOJ) Bureau of Prisons (BOP), and the federal government bears the total cost of incarcerating these criminal aliens. The federal government also reimburses states and localities for a portion of state and local incarceration costs for criminal alien populations that meet the criteria for reimbursement for DOJ's Bureau of Justice Assistance (BJA) State Criminal Alien Assistance Program (SCAAP).[6]

We most recently reported information on criminal alien statistics in 2011.[7] You asked that we update and expand upon the information in that report. Specifically, this chapter provides information on the following:

---

aliens differently, and some entities may report on foreign-born individuals—which could include aliens and U.S. citizens.

[4] Under U.S. immigration law, an alien may be removable on statutory grounds of inadmissibility or deportability. See 8 U.S.C. §§ 1182, 1227, 1229a(c), (e)(2). An alien determined to be removable and not eligible for any requested relief or protection is to be removed pursuant to an administratively final order of removal. 8 C.F.R. § 1241.1.

[5] States and localities include the 50 states, the District of Columbia, counties and cities, and U.S. territories—American Samoa, Guam, Puerto Rico, Northern Mariana Islands, and the U.S. Virgin Islands.

[6] By statute, only the costs of incarceration for "undocumented criminal aliens" are eligible for reimbursement under SCAAP. This statute defines "undocumented criminal alien" for the purposes of SCAAP reimbursement as an alien who has been convicted of a felony or two or more misdemeanors and (1) entered the United States without inspection, (2) was the subject of removal proceedings at the time he or she was taken into custody, or (3) was admitted as a nonimmigrant and at the time he or she was taken into custody had failed to maintain the nonimmigrant status or to comply with the conditions of such status. See 8 U.S.C. § 1231(i)(3)(B). Generally, incarcerated criminal aliens for whom there is no record of admission or who failed to maintain nonimmigrant status after being admitted would be considered eligible for SCAAP reimbursement. An alien with some type of lawful class of admission, such as a permanent resident, at the time of their arrest would generally not meet the SCAAP definition of an undocumented criminal alien unless the alien was in removal proceedings at the time of arrest.

[7] GAO, Criminal Alien Statistics: Information on Incarcerations, Arrests, and Costs, GAO-11-187 (Washington, D.C.: Mar. 24, 2011). In addition, GAO reported on criminal alien statistics in GAO, *Information on Criminal Aliens Incarcerated in Federal and State Prisons and Local Jails*, GAO-05-337R (Washington, D.C.: Apr. 7, 2005); and *Information*

- What is the number and nationality of criminal aliens incarcerated in federal and state prisons and local jails in the United States over the last 6 years?
- What is known about criminal alien arrests and convictions?
- What is known about the costs of incarcerating criminal aliens in the United States?
- What is known about the removability from the United States of criminal aliens incarcerated in federal prison and the experiences of criminal aliens after incarceration in federal prison?

Table 1. Common Terms and Definitions Used in GAO's Report

Term	Definition
Alien	Any person who is not a citizen or national of the United States.
Criminal alien	An alien, with or without lawful status, convicted in the United States of a crime.
State Criminal Alien Assistance Program (SCAAP) criminal alien[a]	A criminal alien—as defined above—incarcerated in a state prison or local jail and for whom a state or locality received federal reimbursement through SCAAP.
Primary offenses (for federal convictions)	When convicted of multiple federal offenses, the primary offense is the one with the longest maximum sentence as determined by the relevant federal agency.
Removability from the United States	Under U.S. immigration law, an alien may be removable on statutory grounds of inadmissibility or deportability. An alien determined to be removable and not eligible for any requested relief or protection is to be removed pursuant to an administrative final order of removal.[b]

Source: GAO. | GAO-18-433.

[a]SCAAP criminal aliens are a subset of all criminal aliens incarcerated in state prisons and local jails because (1) not all states and localities may choose to apply for SCAAP reimbursement and (2) criminal aliens with lawful immigration status who were not the subject of removal proceedings at the time they were taken into custody do not meet the statutory criteria for SCAAP reimbursement. See 8 U.S.C. § 1231(i)(3)(B). Further, to be eligible for reimbursement, the aliens must meet the following criteria: (1) had at least one felony or two misdemeanor convictions for violations of state or local law and (2) were incarcerated for at least 4 consecutive days during the reporting period. For our analyses, "state prisons and local jails" include those in U.S. territories, unless otherwise noted.

[b]See 8 U.S.C. §§ 1182, 1227, 1229a(c), (e)(2); see also 8. C.F.R. § 1241.1.

---

on Certain Illegal Aliens Arrested in the United States, GAO-05-646R (Washington, D.C.: May 9, 2005).

In addition, as you requested, we also describe what is known about certain individuals with international terrorism-related convictions. This information is provided in an appendix.

Table 1 summarizes key terms and definitions used in this chapter.

In general, we analyzed data separately for criminal aliens incarcerated in federal prisons and SCAAP criminal aliens incarcerated in state prisons and local jails—which we refer to as our two study populations.[8] The time periods we analyzed varied for our federal study population compared to our state and local study population because they reflect updates since we last reported on these issues in 2011 and because we used the most recent data available at the time of our analysis.[9] Our federal study population generally includes criminal aliens incarcerated from fiscal years 2011 through 2016.[10] Our state and local study population includes SCAAP criminal aliens incarcerated in state prisons and local jails from fiscal years 2010 through 2015.[11] Overall, our findings are not generalizable to criminal aliens not included in our federal and state and local study populations. However, they provide valuable insights into the criminal aliens incarcerated in the United States. For example, we used SCAAP data because there are no reliable population data on all criminal aliens

---

[8] For our analyses, "state prisons and local jails" include those in U.S. territories, unless otherwise noted.

[9] GAO-11-187.

[10] We used both BOP snapshot (point in time) and inmate-level data to conduct our analyses, as described throughout this report. The BOP inmate-level data included about 198,000 criminal aliens incarcerated in federal prisons from 2011 through 2016. For some of our analyses, we used a smaller subset of the BOP inmate-level data as explained throughout the report. However, when analyzing federal costs to incarcerate criminal aliens in federal prisons, we used BOP snapshot data from fiscal years 2010 through 2015, as these years ensured there were no reporting gaps from our prior report and these were the most recent data available on federal costs for reimbursing states and localities.

[11] We used both SCAAP jurisdiction-level and inmate-level data to conduct our analyses, as described throughout this report. For the SCAAP inmate-level data, we were not able to determine how many unique SCAAP criminal aliens were in the data set, since a SCAAP criminal alien could have more than one incarceration in the same fiscal year. As a result, when reporting on these data for certain analyses, we refer to SCAAP criminal alien incarcerations rather than SCAAP criminal aliens. However, we were able to determine that approximately 533,000 SCAAP inmate-level records had a unique Federal Bureau of Investigation (FBI) number, which we used for certain analyses. For some of our analyses, we used a smaller subset of the SCAAP inmate-level data as explained throughout the report.

incarcerated in every U.S. state prison and local jail.[12] SCAAP provides reliable data on certain criminal aliens incarcerated in state prisons and local jails but does not include (a) criminal aliens incarcerated in states or localities that did not apply for and receive federal reimbursement for costs of incarceration and (b) aliens with lawful immigration status who were not the subject of removal proceedings at the time they were taken into custody.[13] Further, to be eligible for reimbursement, the aliens must meet the definition of "undocumented criminal alien" under the SCAAP statute and the following criteria: (1) had at least one felony or two misdemeanor convictions for violations of state or local law and (2) were incarcerated for at least 4 consecutive days during the reporting period.[14] Thus, our state and local criminal alien data represent only a portion of the total population of criminal aliens incarcerated at the state and local level.

To determine the number and nationalities—based on country of citizenship or country of birth data—of incarcerated criminal aliens, we analyzed BOP data on criminal aliens incarcerated in federal prisons from fiscal years 2011 through 2016 and SCAAP data on SCAAP criminal alien incarcerations in state prisons and local jails from fiscal years 2010 through 2015.[15] BOP obtains country of citizenship data from presentence investigation reports, which may be based on documentation or be

---

[12] In addition to SCAAP data, DOJ's Bureau of Justice Statistics collects data on noncitizens incarcerated in state prisons but these data do not include all states. For example, in 2016, the Bureau of Justice Statistics reported that certain states—including California, which has the highest number of SCAAP criminal aliens—did not report or were unable to report data on the number of noncitizens. U.S. Department of Justice, Bureau of Justice Statistics, *Prisoners in 2015*, (Washington, D.C., Dec. 2016).

[13] SCAAP data represent the number of incarcerations, rather than number of SCAAP criminal aliens, since these aliens could have multiple SCAAP incarcerations in the same fiscal year.

[14] See 8 U.S.C. § 1231(i)(3)(B).

[15] Each year's SCAAP program is based on SCAAP criminal aliens incarcerated during the prior fiscal year, July 1 through June 30. For example, the fiscal year 2016 SCAAP program will include SCAAP criminal aliens that were incarcerated in state prisons and local jails from July 1, 2014 to June 30, 2015. According to DOJ officials, private facilities are not eligible for SCAAP reimbursement, and states and localities are not eligible to apply on their behalf. For the BOP data, we used the average of 12 monthly snapshots to account for possible differences in incarceration numbers month to month for each fiscal year. In addition, we used inmate data to determine the number of unique inmates, versus a snapshot in time, which may include duplicates across fiscal years.

self-reported.[16] SCAAP country of birth data are provided to DOJ by states and localities that participate in SCAAP.[17]

To determine the number and types of offenses for which criminal aliens were arrested and convicted, we analyzed various federal and state data. Specifically, for arrests, we matched a random sample of 500 criminal aliens from our federal study population and 500 SCAAP criminal aliens from the state and local study population to DOJ's FBI database that maintains data from reporting law enforcement agencies across the nation.[18] While the samples we selected for our analyses allowed us to estimate and provide valuable insights about the arrest history of the approximately 197,000 criminal aliens in our federal study population and the approximately 533,000 SCAAP criminal aliens in our state and local study population, our analyses are not generalizable to the arrest history of criminal aliens not in these study populations. We analyzed the arrest histories of criminal aliens in our samples to estimate the number and types of offenses for which criminal aliens in our federal and state and local study populations were arrested/transferred. We defined an arrest/transfer

---

[16] A federal probation officer completes a presentence investigation report after conducting a presentence interview as well as an independent investigation of the offense and the defendant's background. See 18 U.S.C. § 3552. BOP officials stated that BOP's citizenship data could be updated over time as BOP obtains additional information from other sources, such as information from DHS. We did not independently verify citizenship data.

[17] We did not independently verify country of birth of SCAAP criminal aliens incarcerated in state prisons and local jails.

[18] For this analysis, we only included criminal aliens from each study population if they had an FBI number available. This is a unique identifier used by the FBI, as this was needed to match data across databases. As such, the study populations that we projected to for these analyses were smaller than the originating study populations. For example, our federal study population started with approximately 198,000 criminal aliens, about 197,000 of whom had FBI numbers. For this analysis, we therefore included approximately 197,000 criminal aliens in our federal study population. For our state and local study population, there were approximately 533,000 SCAAP records that had a unique FBI number in the SCAAP data set that were included in our study population. Some of the records in our samples of 500 criminal aliens from our federal study population and 500 SCAAP criminal aliens from the state and local study population had to be excluded for various reasons, including invalid FBI numbers. As a result, we analyzed data for 496 criminal aliens in our federal study population and 487 SCAAP criminal aliens in our state and local study population. All percentage estimates presented in this report for these analyses have a margin of error of plus or minus 7 percentage points or fewer. All estimates of the number of arrests/transfers or offenses have a relative error of plus or minus 14 percent of the estimate or less. The 7 percentage point margin of error and 14 percent relative error represent the upper bounds for the estimates included in this report. See appendix I for more details.

as one of these actions occurring on the same day by the same arresting agency. The data did not allow us to distinguish between a new arrest and a transfer from one agency to another; therefore, we refer to these collectively as "arrests/transfers." The criminal aliens in our samples had arrests/transfers that ranged from 1964 through 2017.[19] Because law enforcement entities send arrest information to the FBI on a voluntary basis, FBI data on arrest history may not include all arrests.[20] An arrest does not necessarily result in prosecution or a conviction of all, or any, of the offenses for which an individual is arrested. To determine the primary offenses for which the approximately 198,000 criminal aliens in our federal study population were convicted and incarcerated, we analyzed BOP conviction data for criminal aliens incarcerated in federal prisons from fiscal years 2011 through 2016. In addition, to determine the types of primary offenses for which SCAAP criminal aliens were convicted, we analyzed conviction data from five state prison systems— Arizona, California, Florida, New York, and Texas—from fiscal years 2010 through 2015. We selected these five state prison systems because they had the most SCAAP criminal alien incarcerations in fiscal year 2015. Collectively, these five state prison systems accounted for 64 percent of the SCAAP criminal alien incarcerations in state prisons during fiscal year 2015.[21] They are also the same state prison systems that we analyzed in our 2011 report on criminal aliens.[22] The information obtained from the selected state prison systems is not generalizable to all state prison systems, but provides useful insights about why SCAAP criminal aliens were incarcerated.

---

[19] FBI officials provided all available records on the criminal aliens in our study populations in August and September 2017. Therefore, any additional arrest/transfer data that were added to these records through the end of the calendar year 2017 would not be included in our analysis.

[20] In the month of December 2017, the FBI reported that approximately 23,300 local, state, tribal, federal, and international partners submitted criminal and/or civil electronic submissions to the FBI.

[21] Our analysis included state prison systems that participated in SCAAP and did not include U.S. territories. State prison systems in Arkansas, West Virginia, Vermont, and the District of Columbia did not receive reimbursement for SCAAP criminal aliens incarcerated in fiscal year 2015.

[22] GAO-11-187.

To determine the costs associated with incarcerating criminal aliens in the United States, we obtained and analyzed cost and inmate data from BOP, SCAAP, and states and localities. Specifically, we analyzed costs to (1) the federal government, (2) state prison systems, and (3) selected states and localities. To determine the annual costs to the federal government to incarcerate criminal aliens, for fiscal years 2010 through 2015, we calculated and combined: (1) the estimated costs to incarcerate criminal aliens in federal prisons and (2) the costs to reimburse states and localities for incarcerating SCAAP criminal aliens. To determine the estimated costs associated with incarcerating SCAAP criminal aliens in state prisons, we used a study by DOJ's Bureau of Justice Statistics that estimated state prison expenditures for medical care, food service, and utilities for all 50 state prison systems in 2001.[23] Applying relevant price deflators and SCAAP reimbursement data, we calculated these selected operating costs (medical care, food service, and utilities) for incarcerating SCAAP criminal aliens for each state prison system that sought SCAAP reimbursement from fiscal years 2010 through 2015. While our estimates provide insight into state expenditures to incarcerate SCAAP criminal aliens, our estimates may not represent actual costs if per capita prison expenditures for incarcerating criminal aliens grew at a different rate than the inflation factors that we used for each category. In addition to estimated operating costs for medical care, food service, and utilities, we also used data on correctional officer salaries from SCAAP. To determine the cost of incarcerating SCAAP criminal aliens in selected states and localities, we analyzed cost and SCAAP data for five state prison systems—Arizona, California, Florida, New York, and Texas—and five local jail systems—Maricopa County, Arizona; Orange County, California; Los Angeles County, California; Essex County, New Jersey; and Harris County, Texas. We selected these state prison and local jail systems because they had the most SCAAP criminal alien incarcerations in 2015.[24]

[23] Bureau of Justice Statistics, *State Prison Expenditures, 2001* (Washington, D.C.: June 2004). These were the most recent cost data available for our purposes. Data were not available on local jail systems.

[24] We selected the top six local jail systems, which accounted for 19 percent of all SCAAP criminal alien incarcerations in local jails for fiscal year 2015. However, we could not

To determine what is known about the removability from the United States of criminal aliens incarcerated in federal prison and the experiences of criminal aliens after incarceration in federal prison, we matched data from criminal aliens incarcerated in federal prison with other available data sources using various identifiers, including alien numbers and FBI numbers.[25] Specifically, to analyze the potential removability from the United States of criminal aliens at the time they were incarcerated in federal prison, we matched BOP data on criminal aliens incarcerated from fiscal years 2011 through 2016 with data from ICE's Alien Criminal Response Information Management System (ACRIMe). ICE specialists use ACRIMe to provide an indication of an individual's identity and potential removability to law enforcement partners, at the request of the law enforcement partner. Because a criminal alien's removability from the United States can change over time, we identified the ACRIMe record with a date that was closest to the date the criminal alien first entered federal prison during our period of analysis and used this record for our analysis. To determine what is known about the experiences of criminal aliens after their incarceration in federal prison, we matched data from those criminal aliens who completed a term of incarceration in federal prison from fiscal years 2011 through 2016 with DHS and DOJ data to determine if these criminal aliens were subsequently removed from the United States, reincarcerated, and/or received naturalized citizenship. We compared the dates the criminal alien completed a term of incarceration in federal prison with the dates of encounters with the federal government and/or law enforcement agencies to determine if those encounters took place after the criminal alien completed a term of incarceration in federal prison. For more details on our scope and methodology, see appendix I.

We determined that the data used in each of our analyses were sufficiently reliable for the purposes of this chapter by analyzing available documentation, such as related data dictionaries; interviewing officials

---

estimate total costs for New York City, New York, as was done in the 2011 GAO report. Officials from this locality stated that they no longer apply for SCAAP funds, and they did not provide us an average daily cost per inmate, see GAO-11-187.

[25] An alien number, or alien registration number, is a unique number assigned to a noncitizen's administrative file for tracking purposes.

knowledgeable about the data; conducting electronic tests to identify missing data and anomalies; and following up with officials, as appropriate.

We conducted this performance audit from August 2016 to July 2018 in accordance with generally accepted government auditing standards. Those standards require that we plan and perform the audit to obtain sufficient, appropriate evidence to provide a reasonable basis for our findings and conclusions based on our audit objectives. We believe that the evidence obtained provides a reasonable basis for our findings and conclusions based on our audit objectives.

BACKGROUND

SCAAP Overview and Reimbursement

SCAAP is intended to provide reimbursement to states and localities for a portion of the correctional officer salary costs associated with incarcerating criminal aliens who meet the definition of "undocumented criminal alien" under the SCAAP statute and the following criteria: (1) had at least one felony or two misdemeanor convictions for violations of state or local law and (2) were incarcerated for at least 4 consecutive days during the reporting period.[26] Therefore, SCAAP is not intended to reimburse states and localities for all of the costs associated with incarcerating all criminal aliens.

When applying for SCAAP, states and localities must submit information to DOJ on their state prison or local jail system, such as correctional officer salary costs, as well as records for each criminal alien incarceration that include identifying information, such as the alien's name and date of birth. DOJ reviews the information provided by the states and localities to determine if any of the individual records are invalid for reimbursement because the individual was incarcerated for fewer than the

[26] See 8 U.S.C. § 1231(i)(3)(B).

required 4 days or the incarceration dates were outside the eligibility year, among other things. DOJ also sends the records to DHS, which provides information related to the individual's immigration or citizenship status. Specifically, DOJ uses the information provided by DHS to determine if each individual (1) met the definition of "undocumented criminal alien" (called SCAAP undocumented criminal alien), (2) lacked documentation to confirm the individual's immigration status (called SCAAP unknown criminal aliens), or (3) verified that the individual was a U.S. citizen or an alien who did not meet the definition of "undocumented criminal alien," and therefore was ineligible for reimbursement under SCAAP.[27] According to ICE officials, some of the SCAAP unknown criminal aliens may be in the United States without lawful status.

However, if they have not come into contact with DHS authorities, ICE would not be able to verify their immigration status. For the fiscal year 2016 SCAAP program, which includes records for incarcerations during fiscal year 2015, DOJ determined that 20 percent of the incarcerations for which states or localities submitted a request for SCAAP reimbursement were ineligible for SCAAP reimbursement.[28]

The amount DOJ awards states and localities in SCAAP reimbursements depends on the extent to which DHS data can help verify the alien's status at the time of incarceration and the SCAAP budget, which is appropriated by Congress each year.[29] First, states and localities are eligible to receive reimbursement for 100 percent of the correctional officer salary costs they expended to incarcerate SCAAP undocumented criminal aliens and partial reimbursement for SCAAP unknown criminal aliens, subject to available appropriations. For SCAAP unknown criminal

---

[27] Within DHS, ICE and the U.S. Citizenship and Immigration Services (USCIS) both provide additional data to DOJ, including class of admission, which DOJ uses to help categorize valid records into ineligible, eligible SCAAP undocumented criminal aliens, and eligible SCAAP unknown criminal aliens.

[28] SCAAP records do not represent the number of unique individuals incarcerated since these individuals could be incarcerated in multiple SCAAP states or localities during the same reporting period. As such, these records represent the number of incarcerations— which could include duplicate SCAAP criminal aliens.

[29] Yearly appropriations for SCAAP are as follows (in thousands): fiscal year 2011: $273,352; fiscal year 2012: $240,000; fiscal year 2013: $237,123; fiscal year 2014: $180,000; fiscal year 2015: $185,000; fiscal year 2016: $210,000; and fiscal year 2017: $210,000.

aliens, states are eligible for reimbursement for 85 percent of correctional officer salary costs, counties are eligible for 86 percent reimbursement of the same costs, and cities are eligible for 62 percent reimbursement.[30] Second, depending upon appropriations, the SCAAP program will reimburse a percentage of the states' and localities' eligible correctional officer salary costs.

Arrest Histories and Removability from the United States

Individuals arrested by federal, state, and local law enforcement authorities are generally fingerprinted and their prints may be sent to the FBI.[31] The FBI creates a unique identification number for each individual allowing, among other things, law enforcement to determine an individual's arrest history. The FBI shares these fingerprints with DHS, which enables the sharing of certain biographic, criminal history, and immigration information between the agencies. If there is a match in DHS's system, the FBI sends an inquiry, called an Immigration Alien Query, to ICE's ACRIMe.[32] ICE specialists working on ACRIMe inquiries

---

[30] In 2011, we reported that DOJ's methodology for reimbursing states and localities for SCAAP unknown criminal aliens was based on an analysis that the former Immigration and Naturalization Service conducted in 2000, which analyzed the records of aliens submitted for SCAAP reimbursement in 1997 whose status was at that time unknown. Based upon this analysis, DOJ officials stated that the former Immigration and Naturalization Service determined that 65 percent of those SCAAP unknown criminal aliens submitted for SCAAP reimbursement by states did not have legal status, 60 percent submitted for reimbursement by cities did not have legal status, and 80 percent submitted for reimbursement by counties did not have legal status. See GAO-11-187. Since then, DOJ has assessed its reimbursement methodology and, beginning in fiscal year 2011, made changes to how it reimburses states and localities for SCAAP unknown criminal aliens.

[31] According to FBI officials, the FBI receives fingerprints from federal, state, local, and tribal criminal justice law enforcement agencies and non-criminal justice agencies. These agencies provide fingerprints at their discretion during various points of the criminal justice lifecycle, potentially including during the arrest process and at the time of booking, and for other non-criminal justice purposes, such as at the time an individual files an application for employment. According to FBI officials, the FBI has agreements with 20 states through the National Fingerprint File Program, which allows for automatic file sharing. In addition, the FBI has agreements with some other states to facilitate electronic file sharing. However, some arrest information may not be provided to the FBI.

[32] ACRIMe is an information system used by ICE to receive and respond to inquiries from federal, state, and local law enforcement agencies about individuals arrested, subject to

are to use an individual's biographic and/or biometric identifiers to search various criminal, customs, and immigration databases and determine what is known about an individual's citizenship status or potential removability from the United States. This determination, transmitted in the form of an Immigration Alien Response (IAR), is then sent back to the FBI and shared with the law enforcement authority from which the fingerprints originated.[33] The IAR is also shared with the appropriate ICE field office so that ICE can determine what enforcement action, if any, to take against the individual.[34]

According to ICE officials, the IAR provides a good indication of an individual's potential removability from the United States at a specific point in time; however, ICE's response to these inquiries may not indicate an individual's immigration status with certainty. According to ICE officials, immigration status can be difficult to determine because various agencies maintain information on immigration status in separate databases. ICE officials also stated that the IAR is intended to provide law enforcement agencies with enough information about aliens in their custody that the law enforcement entities can make an informed decision about how to handle the alien, including whether the alien may be removable from the United States. In addition, an individual's immigration status can change over time. For example, an alien with lawful permanent resident status who is convicted of certain crimes may be subject to removal from the United States, which could result in the loss of their lawful permanent resident status.

background checks, or otherwise encountered by those agencies. These inquiries can come directly from the entity or through a match across the FBI and DHS data systems. According to ICE officials, in fiscal year 2017, ICE provided approximately 1.52 million responses to about 13,000 entities.

[33] ICE officials also reported that the IAR also fulfills ICE's statutory requirement to provide immigration status on request.

[34] See 78 Fed. Reg. 10,623 (Feb. 14, 2013).

ICE may remove aliens, including criminal aliens, from the United States who are subject to a final order of removal or following an administrative removability review.[35] ICE may detain these aliens after DHS has encountered these individuals directly or ICE may issue an immigration detainer to federal, state, or local law enforcement agencies for an alien in federal, state, or local custody after the alien is arrested for a criminal offense and if the officer has probable cause to believe that the alien is removable from the United States.[36] For example, a criminal alien incarcerated in federal prison for whom ICE has issued a detainer may be transferred to ICE custody at the time they complete their term of incarceration in federal prison.[37] Individuals removed by ICE may be subject to administrative or criminal consequences on subsequent reentry because of the prior removal.[38]

In some circumstances, an alien may be released from ICE custody due to the decision of an immigration judge, DHS official, or as otherwise required by law. For example, ICE may have limited authority to detain an alien who is subject to a final order of removal for more than 180 days if the individual is unlikely to be able to be removed in the reasonably foreseeable future.[39]

---

[35] Depending on the circumstance, removal orders may be issued by an immigration court, ICE, U.S. Customs and Border Protection, or federal courts.

[36] The immigration detainer provides notification of ICE's intent to take custody of a removable alien in federal, state, or local custody after that alien is released from such custody. U.S. Immigration and Customs Enforcement, *Policy Number 10074.2: Issuance of Immigration Detainers by ICE Immigration Officers.* Effective April 2, 2017.

[37] ICE and BOP coordinate to identify criminal aliens in federal prisons who may be removable, and ICE may initiate their removal proceedings before an immigration judge, through the Institutional Hearing Program—a coordinated effort between DHS and DOJ to identify criminal aliens who are serving federal prison sentences and complete their removal proceedings while they are incarcerated. According to DOJ, this process may allow immigration removal cases to be adjudicated prior to an individual's release from federal prison. In addition, according to BOP officials, an ICE liaison has been working with BOP on a daily basis since November 2016 to assist with identifying foreign-born aliens in BOP custody; processing aliens for removal; transferring aliens who are removable from the custody of BOP to ICE, pursuant to an immigration detainer; and ensuring the Institutional Hearing Program is functioning effectively and efficiently.

[38] See, e.g., 8 U.S.C. § 1182(a)(6).

[39] See *Zadvydas v. Davis*, 533 U.S. 678 (2001); but see *Jennings v. Rodriguez*, 138 S. Ct. 830 (2018).

Criminal Alien Incarcerations Have Declined and Incarcerated Criminal Aliens Were Primarily Citizens from One of Six Countries

The Number of Criminal Aliens Incarcerated in Federal Prisons Decreased from Fiscal Years 2011 through 2016 and More Than 90 Percent Were Citizens from Six Countries

The estimated number of criminal aliens incarcerated in federal prisons decreased 22 percent, from about 50,400 in fiscal year 2011 to about 39,500 in fiscal year 2016—decreasing each year since fiscal year 2013. During this same time, the estimated number of total inmates incarcerated in federal prisons decreased 8 percent, from about 199,100 in fiscal year 2011 to about 184,000 in fiscal year 2016, as shown in Figure 1.[40]

The criminal alien proportion of the total estimated federal inmate population also generally decreased from fiscal years 2011 through 2016.[41] Specifically, criminal aliens accounted for 25 percent of the total federal inmate population in fiscal year 2011 and 21 percent in fiscal year 2016.

From fiscal years 2011 through 2016, there were a total of about 198,000 unique criminal aliens that were incarcerated in federal prison.[42] These criminal aliens accounted for 35 percent of the total number of unique inmates incarcerated from fiscal years 2011 through 2016. As shown in Figure 2, 77 percent of the approximately 198,000 unique

---

[40] We estimated the number of inmates by calculating the average of BOP's 12 monthly snapshot data for each fiscal year. Snapshot data represent a point in time and may not include all inmates that were incarcerated in each year. Data do not include inmates in the witness security program and unsentenced inmates. BOP obtains country of citizenship data from various sources, including presentence investigation reports, which may be based on documentation, such as a birth certificate or immigration documents, or be self-reported. BOP officials stated that country of citizenship could be updated over time as BOP obtains additional information from other sources, such as information from DHS.

[41] The criminal alien proportion of all inmates in federal prisons increased slightly from 25 percent in fiscal year 2011 to 26 percent in fiscal year 2012, then decreased each fiscal year from 2013 through 2016.

[42] The BOP inmate-level data are based on overall population data from fiscal years 2011 through 2016 and not snapshots of a point in time. Data do not include inmates in the witness security program and unsentenced inmates.

criminal aliens incarcerated in federal prisons from fiscal years 2011 through 2016 were citizens of Mexico. Ninety-one percent were citizens of one of six countries, including Mexico.

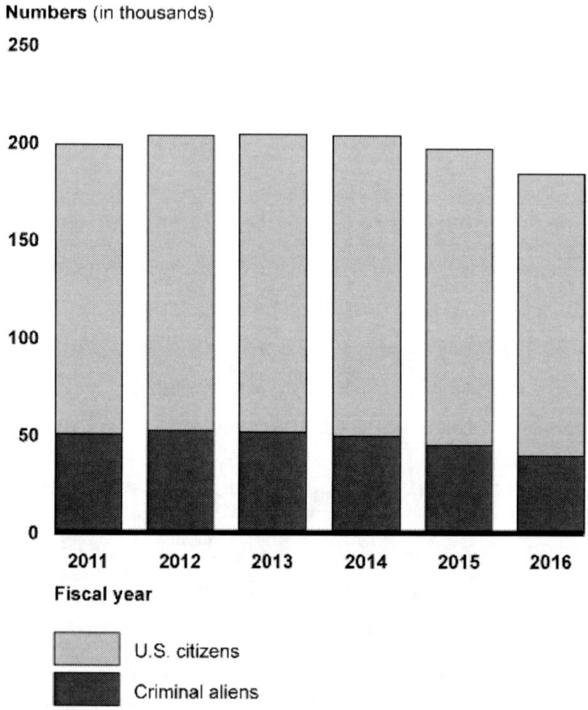

Source: GAO analysis of Bureau of Prisons data. | GAO-18-433.

Note: We estimated the number of inmates by calculating the average of 12 monthly snapshot data for each fiscal year. Snapshot data represent a point in time and may not include all inmates that were incarcerated in each year. Data do not include inmates in the witness security program and unsentenced inmates. These data do not represent unique criminal aliens across the years, as criminal aliens may be incarcerated in more than one fiscal year. The figure above does not include inmates with missing or unknown citizenship data—which represent less than 0.1 percent of the total number of inmates in each fiscal year.

Figure 1. Estimated Number of Individuals, by Citizenship, Incarcerated in Federal Prisons from Fiscal Years 2011 through 2016.

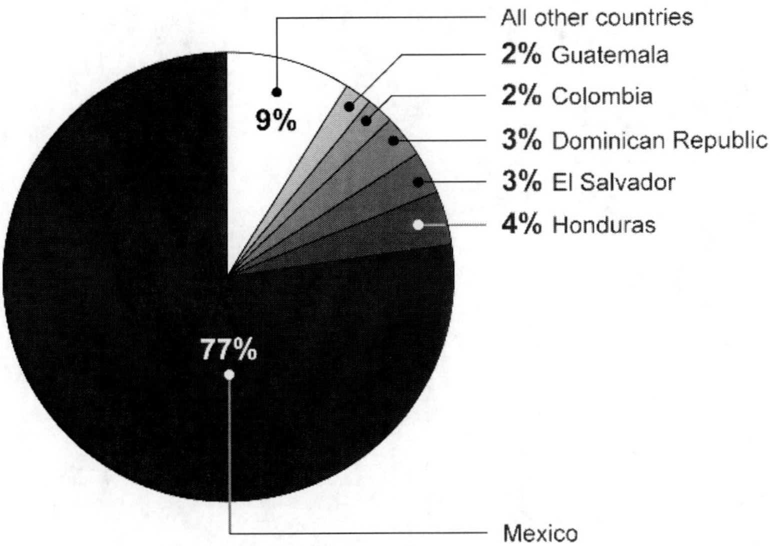

Source: GAO analysis of Bureau of Prisons data. | GAO-18-433.

Figure 2. Percentage of Criminal Aliens Incarcerated in Federal Prisons by Country of Citizenship, Fiscal Years 2011 through 2016.

SCAAP Criminal Alien Incarcerations Decreased 40 Percent from Fiscal Years 2010 through 2015; More Than 75 Percent Were Born in One of Six Countries

From fiscal years 2010 through 2015, the total number of SCAAP criminal alien incarcerations in state prisons and local jails decreased 40 percent from about 282,300 in fiscal year 2010 to about 169,300 in fiscal year 2015—decreasing each year—as shown in Figure 3.[43]

[43] SCAAP data represent the number of incarcerations, rather than number of SCAAP criminal aliens, since aliens could have multiple SCAAP incarcerations in the same fiscal year. As stated previously, SCAAP criminal aliens represent a portion of the total population of criminal aliens incarcerated in state prisons and local jails.

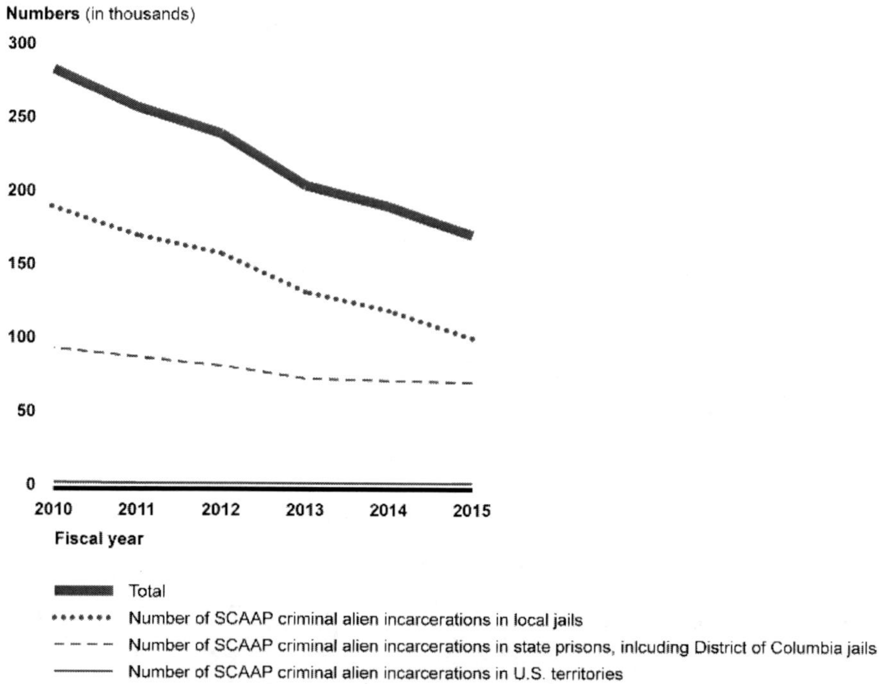

Source: GAO analysis of Bureau of Justice Assistance data. | GAO-18-433.

Note: SCAAP data represent the number of incarcerations, rather than number of SCAAP criminal aliens since these aliens could have multiple SCAAP incarcerations in the same fiscal year. The decrease in the number of SCAAP criminal alien incarcerations can be partially attributed to fewer SCAAP incarcerations. Of the 710 states and localities that participated in SCAAP each of the 6 fiscal years, 80 percent had fewer SCAAP criminal alien incarcerations in fiscal year 2015 compared to fiscal year 2010, and 75 percent had fewer SCAAP criminal alien inmate days in fiscal year 2015 compared to fiscal year 2010. The decrease in the number of SCAAP criminal alien incarcerations can also be partially attributed to general declines in the number of states and localities participating in SCAAP. The number of states and localities participating in SCAAP decreased 13 percent—from 929 in fiscal year 2010 to 811 in fiscal year 2015.

Figure 3. Number of State Criminal Alien Assistance Program (SCAAP) Criminal Alien Incarcerations from Fiscal Years 2010 through 2015.

From fiscal years 2010 through 2015, the number of SCAAP criminal alien incarcerations in state prisons and local jails decreased by 25 and 48 percent, respectively.[44] The decrease in the number of SCAAP criminal alien incarcerations can be partially attributed to fewer SCAAP incarcerations. Of the 710 states and localities that participated in SCAAP each of the 6 fiscal years, 80 percent had fewer SCAAP criminal alien incarcerations in fiscal year 2015 compared to fiscal year 2010, and 75 percent had fewer SCAAP criminal alien inmate days in fiscal year 2015 compared to fiscal year 2010. Further, the decrease can also be partially attributed to general declines in the number of states and localities participating in SCAAP. The number of states and localities participating in SCAAP decreased 13 percent—from 929 in fiscal year 2010 to 811 in fiscal year 2015.[45]

Of the total number of inmate days for all inmates in state prisons and local jails that received SCAAP reimbursements, SCAAP criminal alien incarcerations accounted for 5 percent of all inmate days from fiscal years 2010 through 2015. Of the total number of inmate days for all inmates in U.S. territories that received SCAAP reimbursements, SCAAP criminal alien incarcerations accounted for 3 percent of all days from fiscal years 2010 through 2015.

Further, of the approximately 169,300 SCAAP criminal alien incarcerations in state prisons and local jails nationwide in fiscal year 2015, 66 percent occurred in seven states—California, Texas, Florida, Arizona, New Jersey, New York, and Illinois. In fiscal year 2015, SCAAP criminal aliens accounted for 4 to 10 percent of all inmate days in each of

---

[44] During our selected time period, the following state prison systems did not participate in SCAAP: West Virginia in fiscal year 2010; West Virginia and the District of Columbia in fiscal years 2011 and 2012; Missouri, Tennessee, Wyoming, West Virginia, and the District of Columbia in fiscal year 2013; West Virginia and the District of Columbia in fiscal year 2014; and Arkansas, West Virginia, Vermont, and the District of Columbia in fiscal year 2015. For all 6 years, every state had at least one state prison or local jail system that participated in SCAAP except for the District of Columbia, Vermont, and Wyoming. In addition, not all of the same localities applied in both 2011 and 2015.

[45] Between 46 and 50 state prison systems—including the District of Columbia— participated in SCAAP from fiscal years 2010 through 2015. During this same time period, between 875 to 760 local jail systems participated in SCAAP each fiscal year and between 4 and 5 U.S. territories participated in SCAAP each fiscal year. 710 jurisdictions participated in SCAAP in all 6 fiscal years.

these seven states. See Figure 4 for the number of SCAAP criminal aliens in each state in fiscal year 2015.[46]

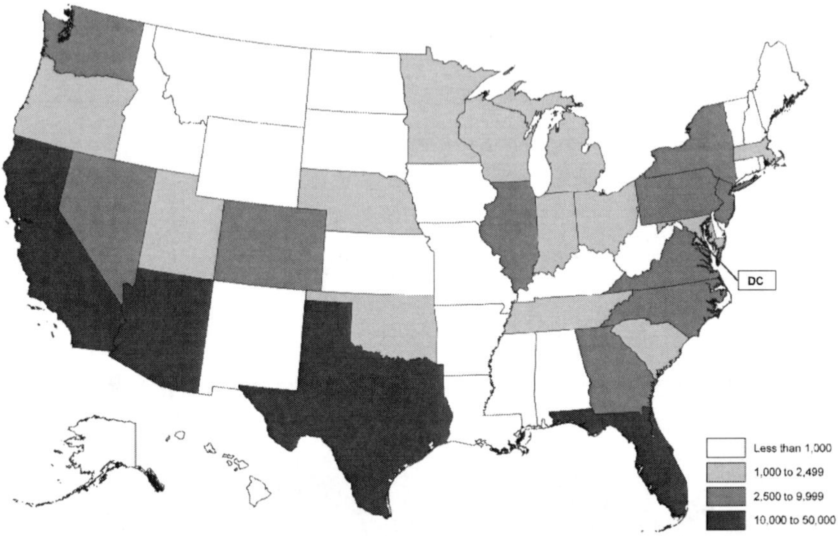

Source: GAO analysis of Bureau of Justice Assistance data. | GAO-18-433.

Note: The above figure includes SCAAP criminal alien incarcerations in state prisons and local jails in each of the states. U.S. territories that received SCAAP reimbursements are not included in the above figure. Vermont and the District of Columbia did not have a state prison or local jail system that applied for reimbursement for criminal aliens incarcerated in fiscal year 2015. SCAAP data represent the number of incarcerations, rather than number of SCAAP criminal aliens, since these aliens could have multiple SCAAP incarcerations in the same fiscal year.

Figure 4. Number of State Criminal Alien Assistance Program (SCAAP) Criminal Alien Incarcerations in Each State, Fiscal Year 2015.

As shown in Figure 5, 57 percent of the SCAAP criminal aliens incarcerated in state prisons and local jails in fiscal year 2015 were born in Mexico. Seventy-six percent were born in one of six countries, including Mexico.[47]

---

[46] See appendix II for data on SCAAP criminal alien incarcerations in state prisons and local jails by state for fiscal years 2010 through 2015.

[47] We did not include about 2,700 incarcerations where the country of birth of the SCAAP criminal alien was unknown. SCAAP country of birth data are provided to DOJ by states

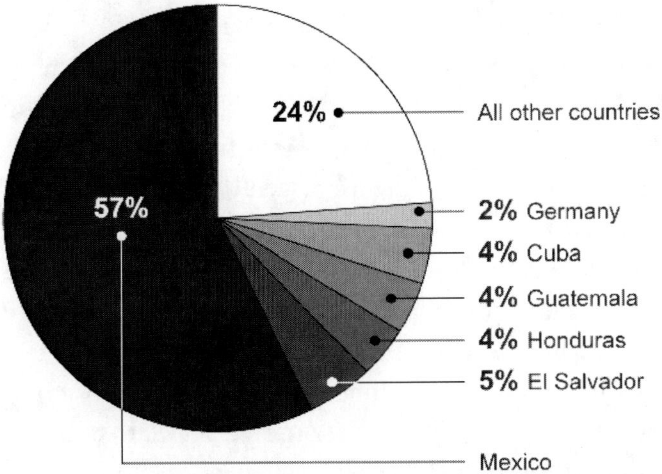

Source: GAO analysis of Bureau of Justice Assistance data. | GAO-18-433.

Note: SCAAP data do not represent the number of unique SCAAP criminal aliens since these aliens could have multiple SCAAP incarcerations during the reporting period.

Figure 5. Percentage of State Criminal Alien Assistance Program (SCAAP) Criminal Alien Incarcerations by Country of Birth, Fiscal Year 2015.

Federal, State, and Local Arrests of Criminal Aliens Were Mostly Related to Immigration and Drug Offenses, As Were Federal Convictions

About Half of Criminal Aliens in Our Study Populations Had One to Five Arrests or Transfers and Most Had at Least One Immigration Offense

Federal Criminal Alien Population

Based on our random sample of 500 criminal aliens incarcerated in federal prisons during fiscal years 2011 through 2016, we estimated that

---

and localities. We did not independently verify country of birth of SCAAP criminal aliens. SCAAP data do not represent the number of unique SCAAP criminal aliens since these aliens could have multiple SCAAP incarcerations during the reporting period.

the about 197,000 criminal aliens in our federal study population were arrested/transferred about 1.4 million times from 1974 through 2017, averaging about 7 arrests/transfers per criminal alien.[48] We estimated that 81 percent of these arrests/transfers occurred from 2000 through 2017. An arrest does not necessarily result in prosecution or a conviction of all, or any, of these offenses.

As shown in Figure 6, about 100,000 (51 percent) of the criminal aliens in our federal study population had one to five arrests/transfers in their arrest history record since 1974.[49]

As shown in Figure 7, we estimated that 65 percent of criminal aliens in our federal study population were arrested/transferred in one of three states—Texas, California, and Arizona.

---

[48] These dates, 1974 through 2017, represent the dates of the oldest and newest arrest/transfer records in our FBI dataset. Since the data did not allow us to determine the difference between a new arrest and a transfer from one arresting agency to another, we are reporting on arrests and transfers together. For this analysis, we only included criminal aliens in the study population if they had an FBI number available, as this was needed to match data across databases. Of the approximately 198,000 criminal aliens incarcerated in federal prisons from fiscal years 2011 through 2016, about 197,000 had FBI numbers. As such, for this analysis, we drew a random sample of 500 criminal aliens from our federal study population of about 197,000 criminal aliens. Some of the records in our sample of 500 criminal aliens had to be excluded for various reasons. As a result, we analyzed data for 496 criminal aliens. While our analyses allowed us to estimate and provide valuable insights about the arrest history of the approximately 197,000 criminal aliens in our federal study population, our analyses are not intended to infer conclusions about the arrest history of criminal aliens not in this study population. Because our sample is drawn from our federal study population and does not include the SCAAP criminal aliens from our state and local study population, results cannot be compared to the results we presented in our 2011 report, see GAO-11-187. Law enforcement entities send arrest information to the FBI on a voluntary basis; as a result, FBI data on arrest history may not include all arrests. All percentage estimates presented in this report have a margin of error of plus or minus 7 percentage points or fewer. All estimates of the number of arrests/transfers or offenses have a relative error of plus or minus 14 percent of the estimate or less. See appendix I for more details on the margin of error for these estimates. See appendix III for the breakdown of arrests/transfers by federal arresting agencies and state and local arresting agencies.

[49] We estimated that about 9,900 (5 percent) of the approximately 197,000 criminal aliens in our federal study population had one arrest/transfer since 1974.

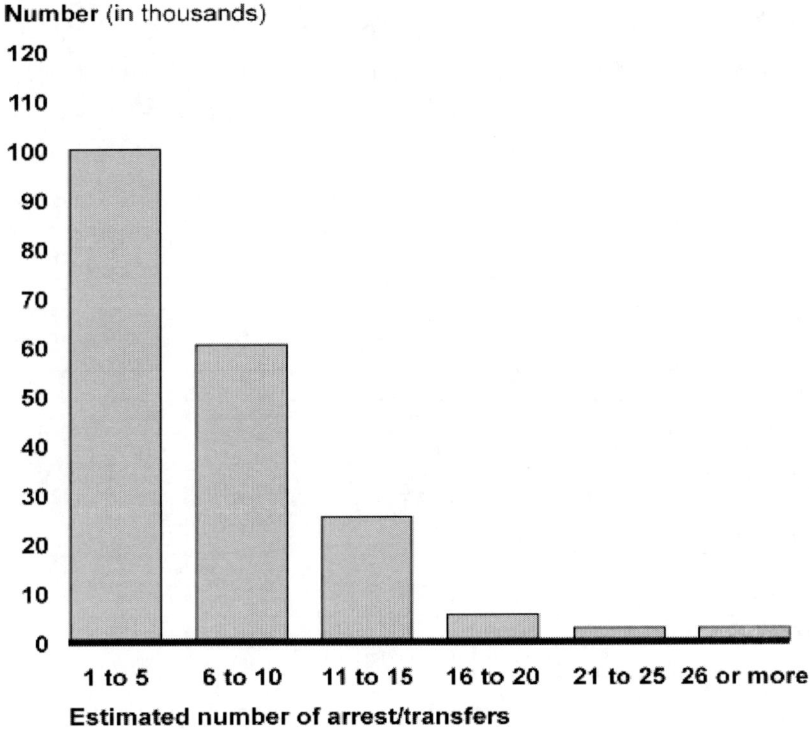

Source: GAO analysis of Federal Bureau of Investigation (FBI) data. | GAO-18-433.

Note: While our analyses allowed us to estimate and provide valuable insights about the arrest history of our study population, which consisted of about 197,000 criminal aliens incarcerated in federal prisons from fiscal years 2011 through 2016 with an FBI number, our analyses are not intended to infer conclusions about the arrest history of criminal aliens not in this study population. Law enforcement entities send arrest information to the FBI on a voluntary basis; as a result, FBI data on arrest history may not include all arrests. The data did not allow us to determine the difference between a new arrest and a transfer from one arresting agency to another; as such, we are reporting on arrests and transfers collectively. All estimates in this figure have a margin of error of +/- 5 percentage points or fewer.

Figure 6. Estimated Number of Arrests/Transfers from 1974 through 2017 per Criminal Alien Incarcerated in Federal Prisons from Fiscal Years 2011 through 2016 who had an FBI Number.

Table 2. Estimated Number and Percent of Attempted or Committed Offenses for Which Criminal Aliens Incarcerated in Federal Prisons from Fiscal Years 2011 through 2016 who had an FBI Number Were Arrested/Transferred from 1974 through 2017

Arrest offense	Estimated number	Estimated percent
Immigration[a]	874,400	42.4
Drugs	336,600	16.3
Traffic violations	204,400	9.9
Obstruction of justice	141,300	6.8
Assault	108,400	5.3
Miscellaneous	74,200	3.6
Larceny/theft	70,300	3.4
Fraud, forgery, and counterfeiting	62,300	3.0
Burglary	44,900	2.2
Weapons violations	44,500	2.2
Motor vehicle theft	19,500	0.9
Sex offenses	13,500	0.7
Disorderly conduct	12,300	0.6
Stolen property	14,300	0.7
Property damage	17,500	0.8
Robbery	13,500	0.7
Homicide	6,000	0.3
Kidnapping	5,600	0.3
Arson	400	< 0.1
Terrorism	400	< 0.1
Total[b]	2,064,100	100

Source: GAO analysis of Federal Bureau of Investigation (FBI) data. | GAO-18-433.

Note: These include offenses associated with arrests or transfers by federal arresting agencies and state and local arresting agencies. Offenses may include an attempt or conspiracy to commit the respective offense. While our analyses allowed us to estimate and provide valuable insights about the arrest history of our study population, which consisted of about 197,000 criminal aliens incarcerated in federal prisons from fiscal years 2011 through 2016 with an FBI number, our analyses are not intended to infer conclusions about the arrest history of criminal aliens not in this study population. Law enforcement entities send arrest information to the FBI on a voluntary basis; as a result, FBI data on arrest history may not include all arrests. The numbers in the table above represent the number of offenses we estimated that criminal aliens were arrested or transferred for; they do not represent the number of times that criminal aliens were arrested or transferred for each offense. All estimates in this table have a margin of error of +/- 3 percentage points or fewer.

[a]Offenses included in our immigration category include both criminal immigration offenses (about 369,200) and civil immigration violations— administrative grounds of removability (about 494,600). For the remaining immigration offenses, the data did not allow us to distinguish whether the offense was criminal or civil.

[b]Numbers may not sum to totals because of rounding. Percentages may not sum to 100 due to rounding.

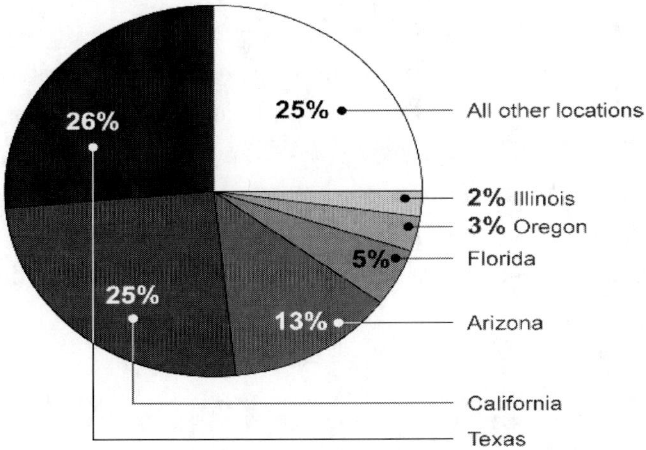

Source: GAO analysis of Federal Bureau of Investigation (FBI) data. | GAO-18-433.

Note: While our analyses allowed us to estimate and provide valuable insights about the arrest history of our study population, which consisted of about 197,000 criminal aliens incarcerated in federal prisons from fiscal years 2011 through 2016 with an FBI number, our analyses are not intended to infer conclusions about the arrest history of criminal aliens not in this study population. Law enforcement entities send arrest information to the FBI on a voluntary basis; as a result, FBI data on arrest history may not include all arrests. The data did not allow us to determine the difference between a new arrest and a transfer from one arresting agency to another; as such, we are reporting on arrests and transfers collectively. All estimates in this figure have a margin of error of +/- 5 percentage points or fewer. Percentages may not add to 100 due to rounding.

Figure 7. Location of Arrests/Transfers from 1974 through 2017 for Criminal Aliens Incarcerated in Federal Prisons from Fiscal Years 2011 through 2016 who had an FBI Number (Estimated Percentage).

We estimated that the approximately 197,000 criminal aliens in our federal study population were arrested/transferred for a total of about 2 million offenses, averaging about 10 offenses per criminal alien.[50] A single

[50] These include offenses associated with arrests or transfers by federal arresting agencies and state and local arresting agencies. For our analysis, (a) multiple counts of the same offense were counted once, (b) duplicate offenses on the same day may be counted more than once if the person was transferred to another agency on the same day and that agency submitted the same offense to the FBI, and (c) duplicate offenses on the same day submitted by the same agency may be counted more than once if there was not enough information to determine that two offenses were the same. Because we selected samples from both our federal study population and our state and local study population, results cannot be compared to the results we presented in our 2011 report, see GAO-11-187. See appendix I

arrest can be for multiple offenses, and being arrested for one or more offenses does not necessarily result in prosecution or a conviction of all, or any, of the offenses for which an individual was arrested. Of the approximately 2 million offenses, we estimated that 42 percent were related to immigration and 26 percent were related to drugs or traffic violations, as shown below in Table 2. Each offense category in the table may include an attempt or conspiracy to commit the respective offense.

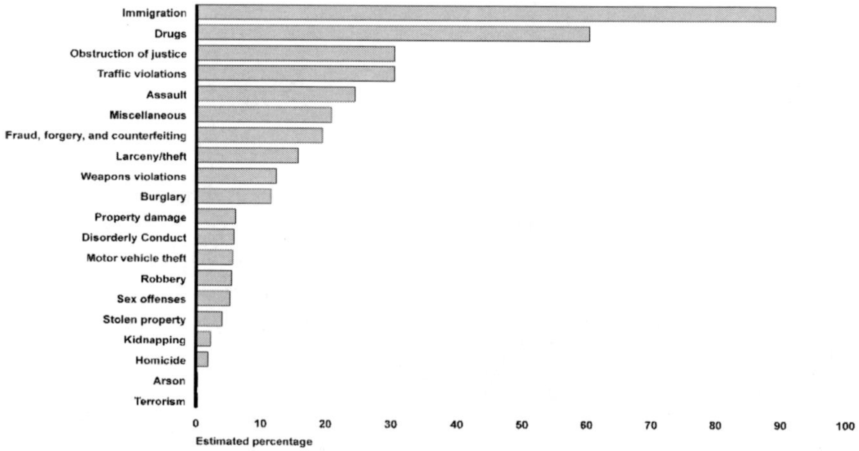

Source: GAO analysis of Federal Bureau of Investigation (FBI) data. | GAO-18-433.

Note: Offenses may include an attempt or conspiracy to commit the respective offense. While our analyses allowed us to estimate and provide valuable insights about the arrest history of our study population, which consisted of about 197,000 criminal aliens incarcerated in federal prisons from fiscal years 2011 through 2016 with an FBI number, our analyses are not intended to infer conclusions about the arrest history of criminal aliens not in this study population. Law enforcement entities send arrest information to the FBI on a voluntary basis; as a result, FBI data on arrest history may not include all arrests. The data did not allow us to determine the difference between a new arrest and a transfer from one arresting agency to another; as such, we are reporting on arrests and transfers collectively. All estimates in this figure have a margin of error of +/- 5 percentage points or fewer.

Figure 8. Estimated Percentage of Criminal Aliens Incarcerated in Federal Prisons from Fiscal Years 2011 through 2016 who had an FBI Number that were Arrested/Transferred At Least Once from 1974 through 2017 by Offense Category, Attempted or Committed.

---

for information on the margin of error for these estimates and a complete description of each of the offense categories.

We estimated that 89 percent of the approximately 197,000 criminal aliens in our federal study population were arrested/transferred at least once for an immigration offense and that 60 percent were arrested/transferred at least once for a drug offense. Figure 8 shows the estimated percentage of criminal aliens arrested/transferred at least once by offense category—which may include an attempt or conspiracy to commit the offense.

SCAAP Criminal Alien Population

Based on our random sample of 500 SCAAP criminal aliens incarcerated in state prisons and local jails, we estimated that our state and local study population of about 533,000 SCAAP criminal aliens were arrested/transferred about 3.5 million times from 1964 through 2017, averaging about 7 arrests/transfers per SCAAP criminal alien.[51] We estimated that 85 percent of these arrests/transfers occurred from 2000 through 2017. An arrest does not necessarily result in prosecution or a conviction of all, or any, of these offenses.

---

[51] These dates, 1964 through 2017, represent the dates of the oldest and newest arrest/transfer records in our FBI dataset. For this analysis, we only included SCAAP criminal aliens in the study population if they had an FBI number available, as this was needed to match data across databases. We determined that approximately 533,000 inmate-level records of SCAAP criminal aliens incarcerated between fiscal years 2010 through 2015 had a unique FBI number. As such, for this analysis, we drew a random sample of 500 criminal aliens from our state and local study population of approximately 533,000 SCAAP criminal aliens with a unique FBI number. Some of the records in our sample of 500 SCAAP criminal aliens had to be excluded for various reasons, including invalid FBI numbers. As a result, we analyzed data for 487 SCAAP criminal aliens. While our analyses allowed us to estimate and provide valuable insights about the arrest history of the about 533,000 SCAAP criminal aliens in our state and local study population, our analyses are not intended to infer conclusions about the arrest history of criminal aliens not in this study population. Because our sample is drawn from our state and local study population and does not include criminal aliens from our federal study population, results cannot be compared to the results we presented in our 2011 report, see GAO-11-187. Law enforcement entities send arrest information to the FBI on a voluntary basis; as a result, FBI data on arrest history may not include all arrests. See appendix I for more details on the margin of error for these estimates. See appendix III for the breakdown of arrests/transfers by federal arresting agencies and state and local arresting agencies. As stated previously, SCAAP criminal aliens represent a portion of the total population of criminal aliens incarcerated in state prisons and local jails.

As shown in Figure 9, about 327,000 (61 percent) of the SCAAP criminal aliens in our state and local study population had one to five arrests/transfers in their arrest history record since 1964.[52]

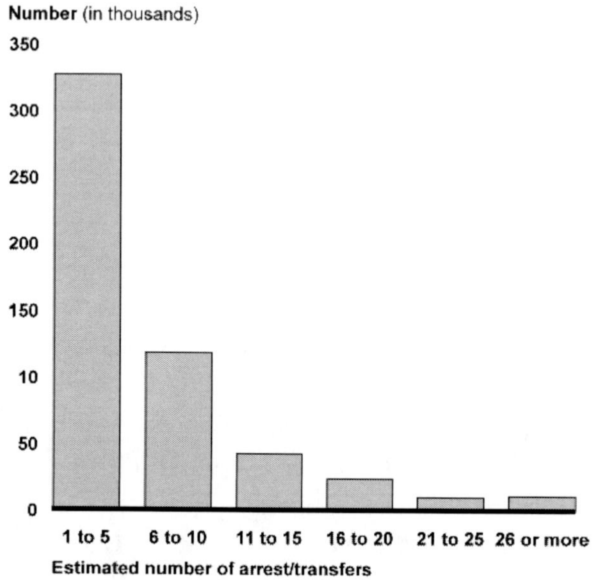

Source: GAO analysis of Federal Bureau of Investigation (FBI) data. | GAO-18-433.

Note: While our analyses allowed us to estimate and provide valuable insights about the arrest history of our study population, which consisted of about 533,000 SCAAP criminal aliens incarcerated in state prisons and local jails from fiscal years 2010 through 2015 with an FBI number, our analyses are not intended to infer conclusions about the arrest history of criminal aliens not in this study population. Law enforcement entities send arrest information to the FBI on a voluntary basis; as a result, FBI data on arrest history may not include all arrests. The data did not allow us to determine the difference between a new arrest and a transfer from one arresting agency to another; as such, we are reporting on arrests and transfers collectively. All estimates in this figure have a margin of error of +/- 5 percentage points or fewer.

Figure 9. Estimated Number of Arrests/Transfers from 1964 through 2017 per State Criminal Alien Assistance Program (SCAAP) Criminal Alien Incarcerated in State Prisons and Local Jails from Fiscal Years 2010 through 2015 who had an FBI Number.

---

52 We estimated that about 41,600 (8 percent) of the approximately 533,000 SCAAP criminal aliens in our state and local study population had one arrest/transfer since 1964.

As shown in Figure 10, we estimated that 62 percent of the SCAAP criminal aliens in our state and local study population were arrested/transferred in one of three states—California, Texas, and Florida.

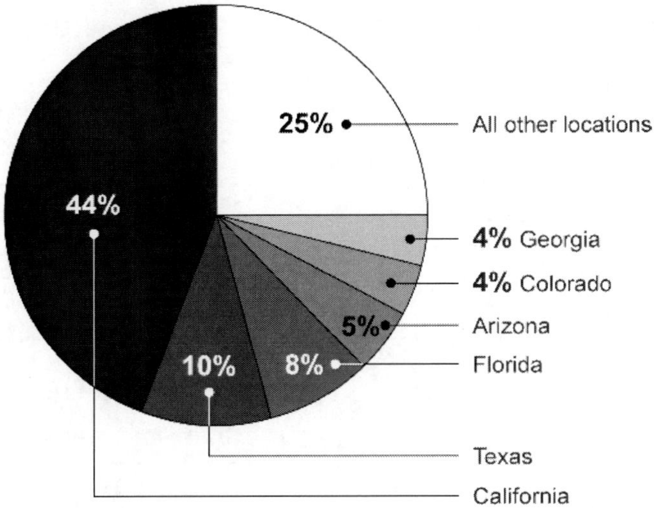

Source: GAO analysis of Federal Bureau of Investigation (FBI) data. | GAO-18-433.

Note: While our analyses allowed us to estimate and provide valuable insights about the arrest history of our study population, which consisted of about 533,000 SCAAP criminal aliens incarcerated in state prisons and local jails from fiscal years 2010 through 2015 with an FBI number, our analyses are not intended to infer conclusions about the arrest history of criminal aliens not in this study population. Law enforcement entities send arrest information to the FBI on a voluntary basis; as a result, FBI data on arrest history may not include all arrests. The data did not allow us to determine the difference between a new arrest and a transfer from one arresting agency to another; as such, we are reporting on arrests and transfers collectively. All estimates in this figure have a margin of error of +/- 6 percentage points or fewer. Percentages may not add to 100 due to rounding.

Figure 10. Location of Arrests/Transfers from 1964 through 2017 for State Criminal Alien Assistance Program (SCAAP) Criminal Aliens Incarcerated in State Prisons and Local Jails from Fiscal Years 2010 through 2015 who had an FBI Number (Estimated Percentage).

Table 3. Estimated Number and Percent of Attempted or Committed Offenses for Which State Criminal Alien Assistance Program (SCAAP) Criminal Aliens Incarcerated in State Prisons and Local Jails from Fiscal Years 2010 through 2015 who had an FBI Number were Arrested/Transferred from 1964 through 2017

Arrest offense	Estimated number	Estimated percent
Traffic violations	1,226,000	22.4
Immigration[a]	852,000	15.6
Drugs	761,200	13.9
Obstruction of justice	665,000	12.2
Assault	397,000	7.3
Larceny/theft	276,700	5.1
Miscellaneous	257,000	4.7
Fraud, forgery, and counterfeiting	200,100	3.7
Burglary	175,000	3.2
Weapons violations	124,700	2.3
Sex offenses	120,300	2.2
Disorderly conduct	90,800	1.7
Motor vehicle theft	90,800	1.7
Stolen property	75,500	1.4
Robbery	54,700	1.0
Property damage	50,300	0.9
Homicide	27,300	0.5
Kidnapping	18,600	0.3
Arson	3,300	0.1
Terrorism	1,100	<0.1
Totalb	5,467,200	100

Source: GAO analysis of Federal Bureau of Investigation (FBI) data. | GAO-18-433.

Note: These include offenses associated with arrests or transfers by federal arresting agencies and state and local arresting agencies. Offenses may include an attempt or conspiracy to commit the respective offense. While our analyses allowed us to estimate and provide valuable insights about the arrest history of our study population, which consisted of about 533,000 SCAAP criminal aliens incarcerated in state prisons and local jails from fiscal years 2010 through 2015 with an FBI number, our analyses are not intended to infer conclusions about the arrest history of criminal aliens not in this study population. Law enforcement entities send arrest information to the FBI on a voluntary basis; as a result, FBI data on arrest history may not include all arrests. The numbers in the table above represent the number of offenses we estimated that SCAAP criminal aliens were arrested or transferred for; they do not represent the number of times that SCAAP criminal aliens were arrested or transferred for each offense. All estimates in this table have a margin of error of +/- 3 percentage points or fewer.

[a]Offenses included in our immigration category included both criminal immigration offenses (about 229,700) and civil immigration violations— administrative grounds of removability (about 616,800). For the remaining immigration offenses, the data did not allow us to distinguish whether the offense was criminal or civil.

[b]Numbers may not sum to totals because of rounding. Percentages may not sum to 100 due to rounding.

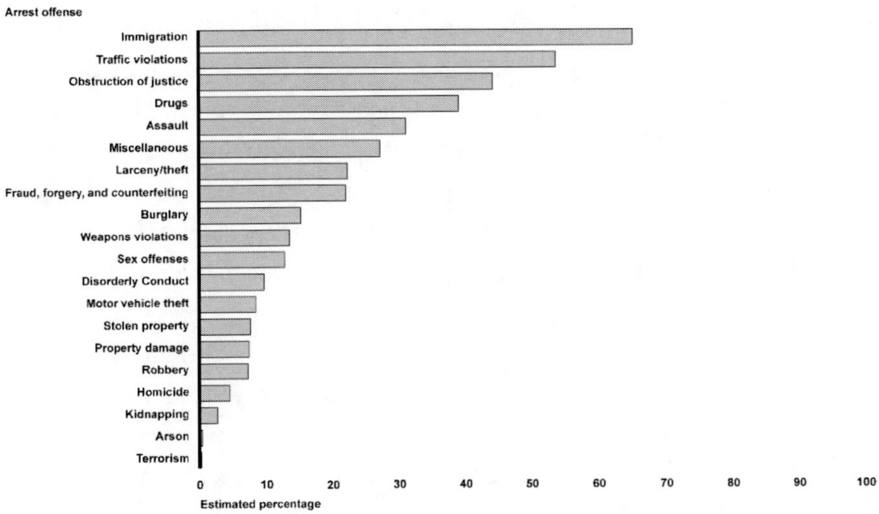

Source: GAO analysis of Federal Bureau of Investigation (FBI) data. | GAO-18-433.

Note: Offenses may include an attempt or conspiracy to commit the respective offense. While our analyses allowed us to estimate and provide valuable insights about the arrest history of our study population, which consisted of about 533,000 SCAAP criminal aliens incarcerated in state prisons and local jails from fiscal years 2010 through 2015 with an FBI number, our analyses are not intended to infer conclusions about the arrest history of criminal aliens not in this study population. Law enforcement entities send arrest information to the FBI on a voluntary basis; as a result, FBI data on arrest history may not include all arrests. The data did not allow us to determine the difference between a new arrest and a transfer from one arresting agency to another; as such, we are reporting on arrests and transfers collectively. All estimates in this figure have a margin of error of +/- 5 percentage points or fewer.

Figure 11. Estimated Percentage of State Criminal Alien Assistance Program (SCAAP) Criminal Aliens Incarcerated in State Prisons and Local Jails from Fiscal Years 2010 through 2015 who had an FBI Number that were Arrested/Transferred At Least Once from 1964 through 2017 by Offense Category, Attempted or Committed.

We estimated that the approximately 533,000 SCAAP criminal aliens in our state and local study population were arrested/transferred for a total of about 5.5 million offenses averaging about 10 offenses per SCAAP criminal alien.[53] A single arrest can be for multiple offenses, and being arrested for one or more offenses does not necessarily result in prosecution or a conviction of all, or any, of the offenses for which an individual was arrested. Of the approximately 5.5 million offenses, we estimated that 52 percent were related to traffic violations, drug offenses, or immigration offenses, as shown below in Table 3. Each offense category in the table may include an attempt or conspiracy to commit the respective offense.

We estimated that 65 percent of the approximately 533,000 SCAAP criminal aliens in our state and local study population were arrested/transferred at least once for an immigration offense and 54 percent were arrested/transferred at least once for a traffic offense. Figure 11 shows the percentage of SCAAP criminal aliens arrested/transferred at least once by offense category—which may include an attempt or conspiracy to commit the offense.

Immigration and Drug Offenses Accounted for the Majority of Federal Convictions of Criminal Aliens, While Selected State Convictions of Criminal Aliens Varied

Because an arrest does not necessarily result in prosecution or a conviction of all, or any, of the offenses for which an individual is arrested, we also analyzed BOP conviction data on criminal aliens incarcerated in federal prisons from fiscal years 2011 through 2016. As shown in Figure 12, the approximately 198,000 unique criminal aliens in our federal study

---

[53] These include offenses associated with arrests or transfers by federal arresting agencies and state and local arresting agencies. See appendix III for the breakdown of offenses for arrests/transfers by federal arresting agencies and state and local arresting agencies. For our analysis, (a) multiple counts of the same offense were counted once, (b) duplicate offenses on the same day may be counted more than once if the person was transferred to another agency on the same day and that agency submitted the same offense to the FBI, and (c) duplicate offenses on the same day submitted by the same agency may be counted more than once if there was not enough information to determine that two offenses were the same. Because we selected samples from both our federal study population and our state and local study population, results cannot be compared to the results we presented in our 2011 report, see GAO-11-187. See appendix I for information on the margin of error for these estimates and a complete description of each of the offense categories.

population that were incarcerated from fiscal years 2011 through 2016 had more than 218,700 primary offenses for which they were convicted. According to BOP data, 92 percent of primary offenses for which these criminal aliens were convicted were related to immigration (63 percent) or drugs (29 percent).[54]

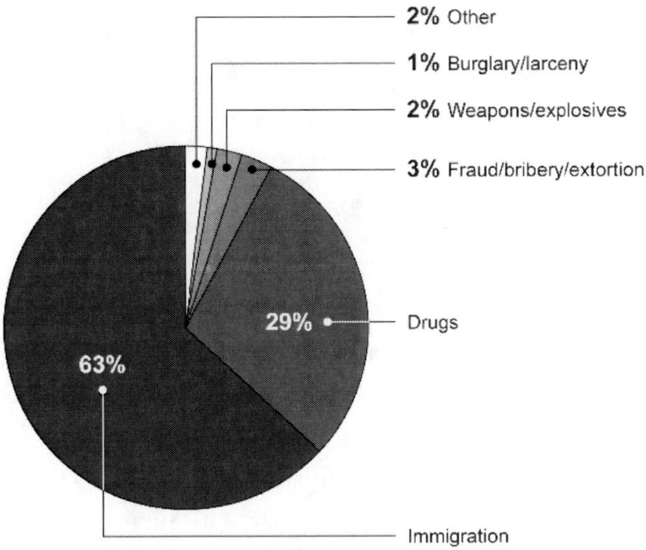

Source: GAO analysis of Bureau of Prisons data. | GAO-18-433.

Note: "Other" includes sex offenses, homicide/aggravated assault, court/corrections, robbery, counterfeit/embezzlement, continuing criminal enterprises, national security, and miscellaneous. If an alien was incarcerated more than once during this time period, each primary offense for which the alien was convicted will be included in these data. As such, there could be more than one primary offense per unique criminal alien. Offenses for which criminal aliens were convicted may include an attempt or conspiracy to commit the respective offense.

Figure 12. Primary Offense Category, Attempted or Committed, for Which the Approximately 198,000 Criminal Aliens Incarcerated in Federal Prisons from Fiscal Years 2011 through 2016 Were Convicted.

[54] If a criminal alien was incarcerated more than once during this time period, each primary offense will be included in these data. As such, there could be more than one primary offense per unique criminal alien. See appendix I for a complete description of each of the offense categories. Offenses may include any attempt or conspiracy to commit the respective offense. BOP data do not include inmates in the witness security program and unsentenced inmates.

In addition to analyzing BOP conviction data for the approximately 198,000 criminal aliens incarcerated in federal prisons from fiscal years 2011 through 2016, we also analyzed U.S. Sentencing Commission data on federal convictions during the same time period to further our understanding of criminal alien convictions. Criminal aliens with convictions in the U.S. Sentencing Commission data may have also been included in our federal study population if they were convicted in federal court and sentenced in fiscal years 2011 through 2016 and also incarcerated in federal prison during this same time period. Specifically, according to U.S. Sentencing Commission data, 91 percent of the approximately 28,000 primary offenses for which criminal aliens were convicted in federal courts and sentenced in fiscal year 2016 were for immigration (66 percent) or drugs (25 percent)—and illegal reentry accounted for 91 percent of the immigration offenses. See appendix IV for additional details on current trends in U.S. Sentencing Commission data on federal convictions. In addition, appendix V provides information on individuals—including U.S. citizens and criminal aliens—convicted as a result of international terrorism investigations.

Since there are no reliable data on criminal aliens incarcerated in all state prisons and local jails, we analyzed conviction data from the five state prison systems that had the largest number of SCAAP criminal alien incarcerations in fiscal year 2015. Primary offenses for which SCAAP criminal aliens incarcerated in fiscal year 2015 were convicted varied across these selected state prison systems:

- Arizona: Drug offenses accounted for 47 percent of the more than 6,300 primary offenses for which SCAAP criminal aliens were convicted
- California: Homicide and sex offenses accounted for about 53 percent of the more than 18,600 primary offenses for which SCAAP criminal aliens were convicted
- Florida: Homicide and sex offenses accounted for 45 percent of the nearly 6,300 primary offenses for which SCAAP criminal aliens were convicted

- New York: Homicide and sex offenses accounted for 49 percent of the nearly 3,400 primary offenses for which SCAAP criminal aliens were convicted
- Texas: Sex, drug, and assault offenses accounted for 52 percent of the nearly 9,600 primary offenses for which SCAAP criminal aliens were convicted

For more information on convictions of SCAAP criminal aliens in these five state prison systems, see appendix VI.

Estimated Federal and State Costs for Incarcerating Criminal Aliens Decreased from Fiscal Years 2010 through 2015

Federal Costs to Incarcerate Criminal Aliens and Federal Reimbursements to States and Localities Decreased from Fiscal Years 2010 through 2015

As previously stated, the federal government incurs costs to incarcerate criminal aliens in federal prisons and to reimburse states and localities for incarcerating SCAAP criminal aliens.[55] From fiscal years 2010 through 2015, these estimated annual total costs decreased 9 percent from about $1.56 billion to about $1.42 billion, as shown in Figure 13.[56]

---

[55] As stated previously, SCAAP criminal aliens represent a portion of the total population of criminal aliens incarcerated in state prisons and local jails.

[56] During this period, costs decreased at an average rate of 1.8 percent per year. Costs associated with incarcerating criminal aliens in federal prisons were estimated based on the average of the 12 monthly population snapshots for each type of BOP institution, and BOP per capita costs. We used the average of the 12 monthly snapshots to account for possible differences in incarceration numbers month to month for each fiscal year. Costs to reimburse states and localities are based on actual costs each year. For data in fiscal year 2016 dollars, see appendix VII.

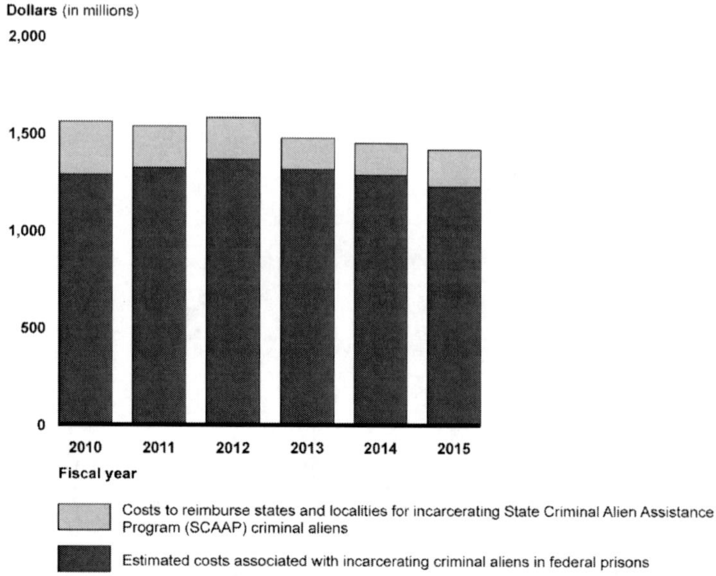

Source: GAO analysis of Bureau of Prisons and Bureau of Justice Assistance data. | GAO-18-433.

Note: Costs to reimburse states and localities are based on actual SCAAP reimbursements each year. Costs associated with incarcerating criminal aliens in federal prisons were estimated based on the average of the 12 monthly population snapshot data for each type of Bureau of Prisons institution and Bureau of Prisons per capita costs.

Figure 13. Federal Costs to Incarcerate Criminal Aliens from Fiscal Years 2010 through 2015.

Most of the costs to the federal government were associated with incarcerating criminal aliens in federal prisons, which accounted for 83 to 89 percent of total costs from fiscal years 2010 through 2015. The estimated cost associated with incarcerating criminal aliens in federal prisons increased from fiscal years 2010 through 2012, and then decreased each year from fiscal years 2012 through 2015. These changes in costs correspond with the changes in the number of criminal aliens incarcerated in federal prison—which also increased from fiscal years 2010 through 2012 and decreased each year thereafter through 2015—while the average daily cost to incarcerate inmates in a federal prison increased each year, as reported by BOP.[57]

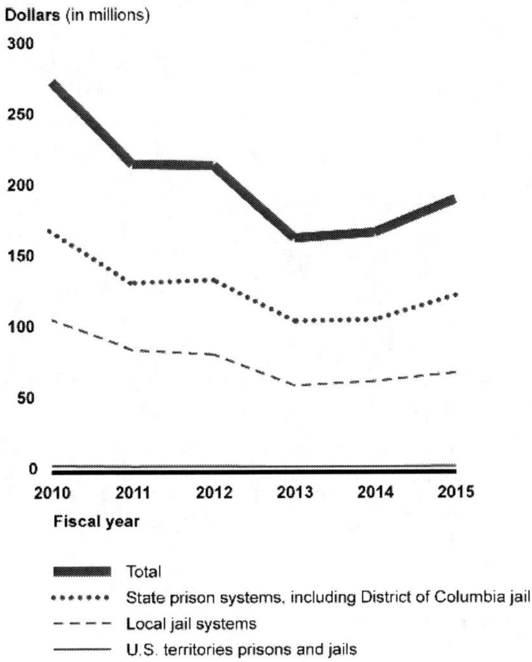

Source: GAO analysis of Bureau of Justice Assistance data. | GAO-18-433.

Figure 14. State Criminal Alien Assistance Program (SCAAP) Reimbursements to States and Localities from Fiscal Years 2010 through 2015.

As shown in Figure 14, the cost to the federal government to reimburse states and localities for incarcerating SCAAP criminal aliens decreased by 31 percent from fiscal years 2010 through 2015, from about $272 million to about $189 million. This decrease reflects lower yearly congressional appropriations.[58] SCAAP reimbursements associated with incarcerating

[57] The average daily cost to incarcerate inmates in federal prison increased 13 percent, from $77.49 in fiscal year 2010 to $87.61 in fiscal year 2015. The average daily cost varies by federal prison facility type. As we reported previously, the number of criminal aliens incarcerated in federal prison increased from fiscal years 2011 to 2012 and then decreased each year since fiscal year 2013.

[58] While the cost to the federal government to reimburse states and localities for incarcerating SCAAP criminal aliens decreased between fiscal years 2010 and 2015, the change in cost varied from year to year. For example, costs decreased by 41 percent from fiscal years 2010

SCAAP criminal aliens in state prisons comprised over 60 percent of SCAAP reimbursements from fiscal years 2010 through 2015, while reimbursements to local jail systems and U.S. territories' prisons and jails comprised the remainder.[59]

Estimated State Selected Operating Costs to Incarcerate SCAAP Criminal Aliens Decreased from Fiscal Years 2010 through 2015; the Federal Government Reimbursed a Portion of These Costs

We estimated that selected annual estimated operating costs—correctional officer salaries, medical care, food service, and utilities—of state prison systems to incarcerate SCAAP criminal aliens decreased 5 percent from fiscal years 2010 through 2015, from about $1.17 billion to about $1.11 billion, as shown in Figure 15.[60] These changes in costs

---

through 2013 before increasing by 17 percent from fiscal years 2013 through 2015. The amount that states and localities were reimbursed is lower than the SCAAP appropriation amount in each fiscal year because some funds were allocated for management and administration, peer review, or other authorized purposes.

[59] For costs in fiscal year 2016 dollars, see appendix VII. The following state prison systems did not receive SCAAP reimbursement: West Virginia in fiscal year 2010; West Virginia and the District of Columbia in fiscal years 2011 and 2012; Missouri, Tennessee, Wyoming, West Virginia, and the District of Columbia in fiscal year 2013; West Virginia and the District of Columbia in fiscal year 2014; and Arkansas, West Virginia, Vermont, and the District of Columbia in fiscal year 2015.

[60] Medical, food and utility expenditures are estimated by using Bureau of Justice Statistics 2001 state prison expenditures by category and adjusting them using the annual Personal Consumption Expenditures Price Index (PCEPI) for each category and the total number of criminal aliens in a given year; per capita expenditures are assumed to grow at the rate of the PCEPI. Correctional officer salaries are based on SCAAP data. While the estimates provide insight into state expenditures to incarcerate SCAAP criminal aliens and make use of actual data on the number of SCAAP criminal aliens and actual 2001 expenditures by state for the selected categories, estimated operating costs might not represent actual expenditures. For instance if per capita prison expenditures are growing at a different rate than the PCEPI, either because of changes in the prison population in the amount consumed or changes in the composition of the services utilized within a category, these costs are not captured by the growth of the PCEPI. For example, based on data we reported in 2017, from fiscal years 2014 to 2015 per capita BOP obligations for health care increased by 9 percent, more than the national medical costs per capita of 5 percent. This is also more than the increase of .6 percent of the PCEPI for medical expenditures during the same period. BOP officials noted that various factors affected the inmate health care costs, including inmates

correspond with a general decrease in the number of SCAAP criminal alien inmate days each year, even though the estimated selected operating costs per inmate generally increased each year.[61]

We also estimated that the total amount that state prison systems expended totaled about $6.7 billion over the 6 years.[62] Of this approximately $6.7 billion, state prison systems were eligible to be reimbursed for about $4.3 billion in correctional officer salaries under SCAAP—the only costs eligible for reimbursement under SCAAP. Based on available appropriations for SCAAP, the amount that the federal government reimbursed state prison systems for incarcerating SCAAP criminal aliens was about $750 million of the approximately $4.3 billion they were eligible to receive, or 17 percent from fiscal years 2010 through 2015. During the same time, we estimated that the federal government's reimbursement, which is for a portion of correctional officer salaries, accounted for 9 percent to 14 percent of these state prison systems' estimated selected operating costs. These selected operating costs might not represent the actual cost of incarceration because our estimates do not incorporate other expenditures beyond the four operating costs. For example, our average daily cost data from selected state prison systems indicate that overall state costs could be higher. See appendix VIII for the estimated costs—based on these average daily costs—and federal reimbursements to selected state prison and local jail systems in fiscal year 2015.

---

entering with poorer health, aging inmates, rising pharmaceutical prices and outside medical services. All of these factors might not apply to the criminal alien prison population; however it points to the fact that there are various factors affecting medical expenditures in prisons that might not be captured by our estimates. See GAO, *Bureau of Prisons: Better Planning and Evaluation Needed to Understand and Control Rising Inmate Health Care Costs*, GAO-17-379 (Washington, D.C.: June 29, 2017).

[61] We estimated—based on annual selected operating costs—that the average daily costs per inmate for state prison systems that received SCAAP reimbursements increased 9 percent, from $43.93 in fiscal year 2010 to $47.69 in fiscal year 2015. These operating costs vary from state to state. The total number of inmate days associated with SCAAP criminal alien incarcerations in state prisons decreased 16 percent, from about 23.7 million in fiscal year 2010 to 19.9 million in fiscal year 2015.

[62] These selected estimated operating costs might not represent actual costs of incarceration. Our estimates do not incorporate other expenditures and actual data from selected state prison systems indicated that their costs tended to be higher than our estimated costs using selected operating costs. For costs in fiscal year 2016 dollars, see appendix VII.

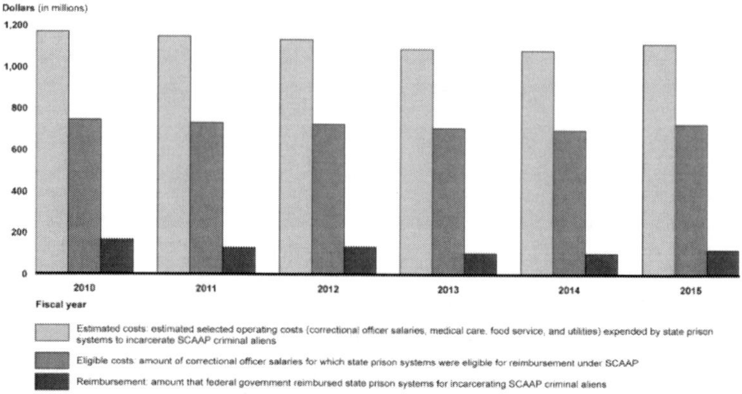

Source: GAO analysis of Bureau of Justice Statistics and Bureau of Justice Assistance data. | GAO-18-433.

Note: SCAAP reimbursement figures may not equal the actual SCAAP awarded amount due to rounding. Our analysis includes those state prison systems—including the District of Columbia jail— that received SCAAP reimbursement in the associated fiscal year the SCAAP criminal aliens were incarcerated. Not all 50 state prison systems received reimbursement each fiscal year and as a result, not all state prison systems may be included in our estimates for each fiscal year.

Figure 15. Estimated Selected Operating Costs to Incarcerate State Criminal Alien Assistance Program (SCAAP) Criminal Aliens in All 50 State Prison Systems from Fiscal Years 2010 through 2015.

Certain Criminal Aliens in Federal Prison Had an Order of Removal or Were Previously Removed from the United States and the Majority Were Removed after Their Incarceration

At the Time of Their Incarceration in Federal Prison, Most Criminal Aliens Who ICE Reviewed Had an Order of Removal or Had Been Previously Removed

To determine the potential removability from the United States of criminal aliens at the time of their incarceration in federal prison, we focused our analysis on those criminal aliens for whom ICE reviewed potential removability using its ACRIMe system within 60 days of the date the alien entered federal prison.[63] According to ICE officials, their response to these inquiries provides a good indication of an individual's potential removability from the United States at a specific point in time. However, as previously noted, ICE's response may not indicate an individual's immigration status with certainty and an individual's immigration status and removability from the United States can change over time.[64] Of the approximately 198,000 criminal aliens incarcerated in federal prison from fiscal years 2011 through 2016, ICE reviewed the potential removability of about 72,600 criminal aliens using its ACRIMe system within 60 days of their incarceration in response to an inquiry from a law enforcement partner.[65] The results of ICE inquiries related to these 72,600 criminal aliens are as follows:

---

[63] As previously discussed, ICE specialists use its ACRIMe system to respond to law enforcement partner inquiries by searching various criminal, customs, and immigration databases to determine what is known about an individual's identity and potential removability from the United States. We used ACRIMe data for this analysis because it provided readily available information about a criminal alien's potential removability as of a particular date. Although ACRIMe is used as a mechanism to share information across agencies, ICE and BOP coordinate in various ways to manage criminal aliens in federal prison. For example, according to BOP officials, an ICE liaison has been working with BOP on a daily basis since November 2016 to assist with identifying foreign-born aliens in BOP custody; processing aliens for removal; transferring aliens who are removable from the custody of BOP to ICE, pursuant to an immigration detainer; and ensuring the Institutional Hearing Program is functioning effectively and efficiently. As previously stated, ICE initiates removal proceedings before an immigration judge, through the Institutional Hearing Program, which may allow immigration removal cases to be adjudicated prior to an individual's release from federal prison.

[64] In some cases, ICE generated multiple ACRIMe responses with different potential removability determinations for the same criminal alien on the same day. ICE officials stated that there are a variety of reasons that ICE specialists might make different potential removability determinations for the same individual on the same day. For instance, analysts are to use the totality of information available in the databases they search, and different analysts may come to different conclusions about the same underlying data. Additionally, available information about an individual's removability from the United States might change, even within the same day.

[65] We excluded about 300 records for which ICE specialists made multiple, different potential removability determinations on the same day within 60 days of the date the alien entered federal prison from this analysis.

- Order of removal or previously removed. About 55,700, or 77 percent, appeared to have a pending order of removal or were previously removed by ICE with no subsequent record of a legal reentry, as indicated by ICE's review of records.[66]
- In removal proceedings. About 5,000, or 7 percent, were in removal proceedings at the time of their incarceration, as indicated by ICE's review of records. In other words, they appeared to have a removal case in process in immigration court.[67]
- Subject to removal. About 11,800, or 16 percent, may have been subject to removal at the time of their incarceration, as indicated by ICE's review of records. For those aliens who may have been subject to removal:
  - About 7,100 criminal aliens may have had some lawful presence or status in the United States, such as being granted a visitor visa, permanent residence, temporary protected status, a border crossing card, or other such authorization to be present in or remain in the United States. If convicted of a felony offense or otherwise in violation of the conditions of their admission, these criminal aliens may be subject to removal from the United States for violating immigration law.
  - About 4,700 criminal aliens did not have a record of admission or a record of other authorized presence in the United States.
- ICE did not have enough information to make an assessment for about 100, or less than 1 percent of the 72,600 criminal aliens it reviewed.[68]

---

[66] See 8 C.F.R. § 1241.8 (stating that "[a]n alien who illegally reenters the United States after having been removed, or having departed voluntarily, while under an order of . . . removal shall be removed from the United States by reinstating the prior order.").

[67] ICE's review of records may indicate that a criminal alien is in removal proceedings. However, ICE does not oversee immigration court proceedings and, as such, may not have complete information about the status of an alien whose case is pending in immigration court.

[68] In these cases, either ICE's records did not have enough information to make an assessment of the identity and potential removability of these criminal aliens or ICE could not find a match for these criminal aliens in the databases searched.

Ninety-Five Percent of Criminal Aliens Were Removed by ICE after Their Incarceration in Federal Prison

Of the 198,000 criminal aliens incarcerated in federal prison from fiscal years 2011 through 2016, approximately 165,700 individuals completed a term of incarceration in federal prison from fiscal years 2011 through 2016.[69] We determined that the approximately 165,700 criminal aliens who completed a term of incarceration in federal prison from fiscal years 2011 through 2016 may have experienced one or more of the following:[70]

ICE Removal

ICE removed about 157,400 of the approximately 165,700 criminal aliens who completed a term of incarceration in federal prison— or 95 percent—at least once after their incarceration in federal prison, as of June 2017.[71] ICE data indicate that the remaining 5 percent of criminal aliens

---

[69] Criminal aliens that did not complete a term of incarceration, as of the end of fiscal year 2016, remained in federal prison and are not included in our analyses. Approximately 280 criminal aliens died while incarcerated in federal prison. These aliens are also not included in our analyses of alien experiences after incarceration in federal prison. Further, approximately 260 criminal aliens who completed a term of incarceration in federal prison from fiscal years 2011 through 2016 either (1) did not have FBI or alien numbers in BOP's records or (2) had duplicate alien numbers in BOP's records. These criminal aliens are also excluded from our analyses of alien experiences after incarceration in federal prison because we could not match their identifiers with data from other agencies.

[70] These outcomes are not mutually exclusive since criminal aliens could have experienced some, none, or all of these after their incarceration in federal prison. For example, an alien could have been removed by ICE after their incarceration in federal prison, then reentered the United States and subsequently become reincarcerated in either federal or state prison or local jail. We matched BOP data on criminal aliens who completed a term of incarceration in federal prison with data from ICE, BJA, and USCIS to determine if these criminal aliens (1) were removed by ICE, (2) were reincarcerated in federal prison or a state prison system or local jail that participated in SCAAP, and/or (3) became naturalized U.S. citizens after their term of incarceration in federal prison. Some criminal aliens may not have had any subsequent encounters with law enforcement after they completed a term of incarceration in federal prison; as a result, there would be no data on these criminal aliens' experiences after their federal prison incarceration.

[71] ICE provided the most recent removal date in its records for each criminal alien who completed a term of incarceration in federal prison from fiscal years 2011 to 2016, as of June 2017. Some aliens may have been removed multiple times between fiscal year 2011 and June 2017, and an alien removed multiple times would only appear in the data with

were not removed by ICE after they completed their term of incarceration in federal prison. For the criminal aliens that ICE did not remove, some did not have an order of removal, some had an order of removal but were not removed by ICE, and some were released from ICE custody, among other potential outcomes, according to ICE data as of June 2017.[72]

The approximately 157,400 criminal aliens ICE removed after their federal prison incarcerations represent a portion of the approximately 1.1 million ICE criminal alien removals from fiscal years 2011 through 2016.[73] ICE's criminal alien removals also include criminal aliens removed after incarcerations in state prisons or local jails, criminal aliens apprehended at or near the border by U.S. Customs and Border Protection and transferred to ICE for removal, and other criminal aliens that ICE apprehends and removes.

We also found that about 3,100 of the approximately 157,400 criminal aliens removed from the United States by ICE after their incarceration in federal prison had a subsequent encounter with ICE in the United States. In other words, these individuals were removed from the United States by ICE after their federal prison incarceration, reentered the United States, and had a subsequent encounter with ICE, as of June 2017.[74]

---

their most recent removal date. ICE matched on BOP data using FBI numbers and/or alien numbers as unique identifiers.

[72] A criminal alien may be released from ICE custody if there is no significant likelihood that they could be removed in the reasonably foreseeable future—for example, some countries will not accept the repatriation of criminal aliens from the United States. ICE has limited authority to indefinitely detain a criminal alien who has a final order of removal. See *Zadvydas v. Davis*, 533 U.S. 678 (2001); but see *Jennings v. Rodriguez*, 138 S. Ct. 830 (2018). Data are available for those criminal aliens ICE encountered after their federal prison incarceration.

[73] For more information about ICE removals from fiscal years 2011 through 2016, see appendix IX. ICE defines a criminal alien as an alien convicted of a crime, either within or outside of the United States. The 1.1 million ICE criminal alien removals from fiscal years 2011 through 2016 are those aliens who meet ICE's definition of a criminal alien. In all other places in this report, a criminal alien is defined as an alien convicted of a crime while in the United States.

[74] Subsequent encounters with ICE include ICE arrests, ICE releases, and admissions to ICE detention facilities, also known as book-ins. ICE provided the most recent date in its records for each type of encounter for each criminal alien who completed a term of incarceration in federal prison, as of June 2017. Some aliens may have had more than one ICE arrest, book-in, removal, or release. As a result, the available data does not allow us to report on the sequence of all criminal alien encounters with ICE after their incarceration in federal prison.

Federal Reincarceration

About 19,300 of the approximately 165,700 criminal aliens (12 percent) who completed a term of incarceration in federal prison from fiscal years 2011 through 2016 were subsequently incarcerated in federal prison at least once, as of the end of fiscal year 2016, as shown in Table 4.

Table 4. Federal Prison Reincarcerations of Criminal Aliens, Fiscal Years 2011 through 2016

Number of federal prison reincarcerations	Number of criminal aliens	Percent
0	146,500	88.4
1	16,700	10.1
2	2,200	1.3
3	300	0.2
4 or more	<100	<0.1
Total[a]	165,700	100

Source: GAO analysis of Bureau of Prisons data. | GAO-18-433.

[a]Numbers may not sum to totals due to rounding. Percentages may not add to 100 due to rounding.

The remaining 88 percent of criminal aliens who completed a term of incarceration in federal prison from fiscal years 2011 through 2016 were not reincarcerated in federal prison, as of the end of fiscal year 2016.

State or Local Reincarceration

We identified about 5,500 of the approximately 165,700 criminal aliens who completed a term of incarceration in federal prison from fiscal years 2011 through 2016 who were subsequently incarcerated in a state prison or local jail system that participated in SCAAP.[75] We did not identify any of the remaining approximately 160,200 criminal aliens who completed a term of incarceration in federal prison in the SCAAP data we reviewed through fiscal year 2015.[76]

---

[75] This number only includes criminal aliens who were subsequently incarcerated (on the same day as or after the day they completed their term of incarceration in federal prison) in a state prison or local jail that applied for SCAAP reimbursement from fiscal years 2010 through 2015.

[76] This analysis includes those criminal aliens for whom we were able to confirm a subsequent incarceration in a state prison or local jail that requested SCAAP reimbursement after their federal prison incarceration. Approximately 533,000 SCAAPeligible unique inmate-level

Naturalization

We identified 16 of the about 165,700 criminal aliens who completed a term of incarceration in federal prison from fiscal years 2011 through 2016 who subsequently became naturalized U.S. citizens, as of June 2017.[77] To be eligible for naturalization, an applicant must generally have 5 years of continuous residence after being lawfully admitted as a permanent resident. Additionally, for the 5 years immediately preceding the application for naturalization, an applicant cannot have been incarcerated for more than 180 days, have been convicted of one or more crimes involving moral turpitude, or have been convicted of a number of other crimes. Certain crimes are permanent bars to naturalization.[78] Of the 16 individuals we identified who became naturalized citizens after their incarceration in federal prison, 7 had been convicted of drug offenses. The remaining 9 individuals had various primary conviction offenses, including: burglary/larceny (3 individuals), fraud/bribery/extortion (2 individuals), immigration (2 individuals), counterfeit/embezzlement (1 individual), and court/corrections (1 individual).

U.S. Citizenship and Immigration Services (USCIS), a component agency within DHS, determined that all 16 of the individuals we identified who received naturalized citizenship after their incarceration in federal prison (1) demonstrated good moral character for the statutory period, (2) completed and resolved all background checks with appropriate law enforcement agencies, and (3) did not demonstrate national security concerns, public safety concerns, or other grounds of inadmissibility.

---

records we analyzed contained an FBI number, which we used to match SCAAP data with BOP's data on criminal aliens. Due to data limitations, we could not match to the remaining SCAAP records or to incarceration records for jurisdictions that did not receive SCAAP reimbursement.

[77] To identify these aliens, we matched BOP and USCIS data using multiple identifiers— alien number, name, and date of birth. We determined that these 16 aliens matched on all identifiers we analyzed. The 16 aliens that we matched represent a minimum number of those who naturalized after their federal prison incarceration during this time period.

[78] Certain convictions, such as convictions for offenses entered on or after November 29, 1990, that fall under the definition of aggravated felony under 8 U.S.C. § 1101(a)(43), act as a permanent bar to good moral character and, as a result, naturalization, regardless of when the conviction occurred.

AGENCY COMMENTS AND THIRD-PARTY VIEWS

We provided a draft of this chapter to DHS and DOJ for their review and comment. We also provided a draft of relevant portions of this chapter to the U.S. Sentencing Commission, Arizona Department of Corrections, California Department of Corrections and Rehabilitation, Florida Department of Corrections, New York State Department of Corrections and Community Supervision, Texas Department of Criminal Justice, Essex County Department of Corrections, Los Angeles County Sheriff's Department, Maricopa County Sheriff's Office, Orange County Sheriff's Department, New York City Mayor's Office of Criminal Justice, and Harris County Sheriff's Office for their review and comment. In its comments, which are reprinted in Appendix X, DOJ noted that we appropriately qualify the data presented on SCAAP, but expressed concern that the SCAAP data may be misinterpreted or misunderstood by readers of our report. We believe that throughout the report, we have appropriately caveated the data we present on SCAAP and the costs associated with the program and therefore have made no changes based on DOJ's comments in their letter. DHS, DOJ, the U.S. Sentencing Commission, Texas Department of Criminal Justice, California Department of Corrections and Rehabilitation, and New York City Mayor's Office of Criminal Justice also provided technical comments, which we incorporated, as appropriate. Arizona Department of Corrections, Florida Department of Corrections, New York State Department of Corrections and Community Supervision, Essex County Department of Corrections, Los Angeles County Sheriff's Department, Maricopa County Sheriff's Office, Orange County Sheriff's Department, and Harris County Sheriff's Office did not have comments.

Gretta L. Goodwin
Director, Homeland Security and Justice Issues

APPENDIX I: OBJECTIVES, SCOPE, AND METHODOLOGY

This appendix provides additional details on our objectives, scope, and methodology. Specifically, our objectives were to provide information on the following:

- the number and nationality of criminal aliens incarcerated in federal and state prisons and local jails in the United States over the last 6 years;
- criminal alien arrests and convictions;
- the costs of incarcerating criminal aliens in the United States; and
- the removability from the United States of criminal aliens incarcerated in federal prison and experiences of criminal aliens after incarceration in federal prison.

In general, we analyzed data separately for criminal aliens incarcerated in federal prisons and State Criminal Alien Assistance Program (SCAAP) criminal aliens incarcerated in state prisons and local jails—which we refer to as our two study populations.[79] The time periods we analyzed varied for our federal study population compared to our state and local study population because they reflect updates since we last reported on these issues in 2011 and because we used the most recent data available at the time of our analysis.[80] Our federal study population generally includes criminal aliens incarcerated in federal prisons from fiscal years 2011 through 2016.[81] For our federal study population, we used both snapshot (point in time) and inmate-level data from the Department of Justice's (DOJ) Bureau of Prisons (BOP) to conduct our analyses, as described

---

[79] For our analyses, "state prisons and local jails" include those in U.S. territories, unless otherwise noted.

[80] GAO, *Criminal Alien Statistics: Information on Incarcerations, Arrests, and Costs,* GAO-11-187 (Washington, D.C.: Mar. 24, 2011).

[81] However, when analyzing federal costs to incarcerate criminal aliens in federal prisons, we used BOP snapshot data from fiscal years 2010 through 2015, as this ensured there were no reporting gaps from our prior report and these were the most recent data available on federal costs for reimbursing states and localities.

throughout this chapter. The BOP inmate-level data included about 198,000 criminal aliens incarcerated in federal prisons from 2011 through 2016.[82] Our state and local study population includes SCAAP criminal aliens incarcerated in state prisons and local jails from fiscal years 2010 through 2015. For our state and local study population, we used DOJ Bureau of Justice Assistance (BJA) SCAAP jurisdiction-level and inmate-level data to conduct our analyses, as described throughout this chapter.[83] For the SCAAP inmate-level data, we were not able to determine how many unique SCAAP criminal aliens were in the data set, since a SCAAP criminal alien could have more than one incarceration in the same fiscal year. As a result, when reporting on these data for certain analyses, we refer to SCAAP criminal alien incarcerations rather than SCAAP criminal aliens. However, we were able to determine that approximately 533,000 SCAAP inmate-level records that had a unique Federal Bureau of Investigation (FBI) number, which we used for certain analyses. Overall, our findings are not generalizable to criminal aliens not included in our federal and state and local study populations. However, they provide valuable insights into the criminal aliens incarcerated in the United States. For example, we used SCAAP data because there are no reliable population data on all criminal aliens incarcerated in every U.S. state prison and local jail.[84] SCAAP provides reliable data on certain criminal aliens incarcerated in state prisons and local jails but does not include (a) aliens incarcerated in states or localities that did not apply for and receive federal reimbursement for costs of incarceration and (b) aliens with lawful immigration status who were not the subject of removal proceedings at the time they were taken into custody. Further, to be eligible for reimbursement, the aliens must meet the definition of "undocumented

---

[82] For some of our analyses, we use a smaller subset of the BOP inmate-level data as explained throughout the report.

[83] For some of our analyses, we used a smaller subset of the SCAAP inmate-level data as explained throughout the report.

[84] In addition to SCAAP data, DOJ's Bureau of Justice Statistics collects data on noncitizens incarcerated in state prisons but these data do not include all states. For example, in 2016, the Bureau of Justice Statistics reported that certain states—including California, which has the highest number of SCAAP criminal aliens—did not report or were unable to report data on the number of noncitizens. U.S. Department of Justice, Bureau of Justice Statistics, *Prisoners in 2015*, (Washington, D.C., Dec. 2016).

criminal alien" under the SCAAP statute and the following criteria: (1) had at least one felony or two misdemeanor convictions for violations of state or local law and (2) were incarcerated for at least 4 consecutive days during the reporting period.[85] Thus, our state and local criminal alien data represent only a portion of the total population of criminal aliens incarcerated at the state and local level.

Number and Nationality of Criminal Aliens

To determine the number and nationalities—based on country of citizenship or country of birth data—of incarcerated criminal aliens, we analyzed BOP data on criminal aliens incarcerated in federal prisons from fiscal years 2011 through 2016 and SCAAP data on SCAAP criminal alien incarcerations in state prisons and local jails from fiscal years 2010 through 2015.[86] Specifically, we analyzed BOP snapshot and inmate-level data from BOP's inmate tracking database, the SENTRY Inmate Management System. To estimate the number and proportion of aliens incarcerated in federal prisons compared to the total inmate population at a point in time each fiscal year, we analyzed BOP data and calculated the average of its 12 monthly population snapshots.[87] In addition, we analyzed BOP inmate-level data to determine the number of unique criminal aliens that were incarcerated in federal prisons during this time frame and the country of citizenship of these aliens.[88] BOP obtains country of citizenship

---

[85] See 8 U.S.C. § 1231(i)(3)(B).

[86] We did not examine the extent to which policy changes may have affected the number of individuals incarcerated in prisons and jails. Each year's SCAAP program is based on SCAAP criminal aliens incarcerated during the prior fiscal year, July 1 through June 30. For example, the fiscal year 2016 SCAAP program will include SCAAP criminal aliens that were incarcerated in state prisons and local jails from July 1, 2014 to June 30, 2015. According to DOJ officials, private facilities are not eligible for SCAAP reimbursement, and states and localities are not eligible to apply on their behalf.

[87] We determined the average of 12 monthly snapshots to account for possible differences in incarceration numbers month to month for each fiscal year. BOP snapshot data do not include inmates in the witness security program and unsentenced inmates. Snapshot data represent a point in time and may not include all inmates that were incarcerated in each month.

[88] BOP data do not include inmates in the witness security program and unsentenced inmates.

data from presentence investigation reports, which may be based on documentation or be self-reported.[89]

For state prisons and local jails, we analyzed SCAAP jurisdiction-level data from SCAAP for fiscal years 2011 through 2016—which includes inmates incarcerated from fiscal years 2010 through 2015—to determine the number of SCAAP criminal alien incarcerations.[90] We compared the number of SCAAP inmate days to the number of total inmate days in these state prisons and local jails to determine the proportion of inmate days that were attributed to SCAAP criminal aliens. In addition, we analyzed SCAAP inmate-level records to determine the country of birth of SCAAP criminal aliens for the most recent fiscal year in which data were available—2015. SCAAP country of birth data are provided to DOJ by states and localities that participate in SCAAP.[91]

Criminal Alien Arrests and Convictions

To determine the number and types of offenses for which criminal aliens were arrested and convicted, we analyzed various federal, state, and local data.[92] Specifically, for arrests, we selected a generalizable random sample of 500 criminal aliens from our federal study population of about 197,000 criminal aliens and a generalizable random sample of 500 SCAAP criminal aliens from the state and local study population of about 533,000 SCAAP criminal aliens. We then matched these 1,000 selected criminal aliens to DOJ's FBI Next Generation Identification (NGI) database.[93] The

---

[89] A federal probation officer completes a presentence investigation report after conducting a presentence interview as well as an independent investigation of the offense and the defendant's background. See 18 U.S.C. § 3552. BOP officials stated that BOP's citizenship data could be updated over time as BOP obtains additional information from other sources, such as information from DHS. We did not independently verify citizenship data.

[90] SCAAP data represent the number of incarcerations, rather than number of SCAAP criminal aliens, since these aliens could have multiple SCAAP incarcerations in the same fiscal year.

[91] We did not independently verify country of birth of SCAAP criminal aliens.

[92] We did not examine the extent to which policy changes may have affected the number of individuals arrested or convicted of crimes.

[93] For this analysis, we only included criminal aliens from each study population if they had an FBI number available. This is a unique identifier used by the FBI, as this was needed to

NGI database maintains data from reporting law enforcement agencies across the nation.[94] The FBI record we obtained for each criminal alien contained the FBI number, name, dates of arrest, the arresting agency, location of the arrest, and a description of each arrest offense.[95] To ensure the data were appropriate for our purposes we, among other things, categorized arresting agencies into federal arresting agencies, such as U.S. Immigration and Customs Enforcement (ICE); state and local arresting agencies, such as state police departments; and non-arresting agencies, such as prisons and jails. While the samples we selected for our analyses allowed us to estimate and provide valuable insights about the arrest history of the approximately 197,000 criminal aliens in our federal study population and the approximately 533,000 SCAAP criminal aliens in our state and local study population, our analyses are not generalizable to the arrest history of criminal aliens not in these study populations.

We subsequently used these data to estimate the number and types of offenses for which criminal aliens in our study populations were arrested/transferred. We defined an arrest/transfer as one of these actions occurring on the same day by the same arresting agency. Because the data did not allow us to distinguish a new arrest from a transfer between one arresting agency and another, we refer to these collectively as

---

match data across databases. As such, the study populations that we projected to for these analyses were smaller than the originating study populations. For example, our federal study population started with approximately 198,000 criminal aliens and about 197,000 had FBI numbers. For this analysis, we included approximately 197,000 criminal aliens in our federal study population. For our state and local study population, there were approximately 533,000 SCAAP records that had a unique FBI number in the SCAAP data set that were included in our study population. Some of the records in our samples of 500 criminal aliens from our federal study population and 500 SCAAP criminal aliens from the state and local study population had to be excluded for various reasons, including invalid FBI numbers. As a result, we analyzed data for 496 criminal aliens in our federal study population and 487 SCCAP criminal aliens in our state and local study population.

[94] The FBI replaced its Integrated Automated Fingerprint Identification System—a national, computerized system for storing, comparing, and exchanging fingerprint data in a digital format –with NGI. NGI is to include fingerprint data from Integrated Automated Fingerprint Identification System and biographic data, and is to also provide new functionality and improve existing capabilities by incorporating advancements in biometrics.

[95] We took steps to ensure that the matches made using the FBI number generally resulted in quality matches of the same individuals across data systems. For example, we reviewed name matches across the records.

"arrests/transfers." We analyzed FBI records for both our samples to estimate the following for our federal and state and local study populations: (a) number of arrests/transfers and (b) number and types of arrest offenses. Because law enforcement entities send arrest information to the FBI on a voluntary basis, FBI data on arrest history may not include all arrests. In the month of December 2017, the FBI reported that approximately 23,300 local, state, tribal, federal, and international partners submitted criminal and/or civil electronic submissions to NGI. The criminal aliens in our samples had arrests/transfers that ranged from 1964 through 2017.[96] We categorized arrest offenses—which includes attempts or conspiracies to commit each of the respective offenses—as shown in Table 5.

Because our estimates regarding criminal alien arrests/transfers and offenses are based on a random sample, each estimate we report has a measurable margin of error due to sampling. For this chapter, the margins of error are calculated based on a 95 percent confidence level.[97] All percentage estimates presented in this chapter for these analyses have a margin of error of plus or minus 7 percentage points or fewer. All estimates of the number of arrests/transfers or offenses have a relative error of plus or minus 14 percent of the estimate or less. The 7 percentage point margin of error and 14 percent relative error represent upper bounds for the estimates included in this chapter.

Since an arrest does not necessarily result in prosecution or a conviction of all, or any, of the offenses for which an individual is arrested, we also analyzed conviction data. To determine the primary offenses for which the approximately 198,000 criminal aliens in our federal study population were convicted and incarcerated, we analyzed BOP inmate-level data on convictions from BOP's SENTRY database for all criminal

---

[96] FBI officials provided all available records on the criminal aliens in our study populations in August and September 2017. Therefore, any additional arrest/transfer data that were added to these records through the end of the calendar year 2017 would not be included in our analysis.

[97] The margin of error surrounding an estimate is expressed as (1) a number of percentage points higher or lower than the percentage estimate, (2) a percent higher or lower than the estimated number, or (3) the entire range the margin of error covers, which is referred to as a confidence interval. Margins of error are calculated based on a certain confidence level.

aliens incarcerated in federal prisons from fiscal years 2011 through 2016. Table 6 describes the BOP primary offense categories for convictions.

Table 5. Arrest Offense Categories, Attempted and Committed

Offense category	Category includes
Arson	Arson, reckless burning, and possession of arson materials
Assault	Assault, battery, assault with a deadly weapon, endangerment, and threats
Burglary	Breaking and entering, burglary, and possession of burglary tools
Disorderly Conduct	Disturbing the peace, fighting, intoxication, public nuisance, and disorderly conduct
Drugs	Use/under the influence, possession, possession with intent to distribute, sales, manufacturing, transporting, and possession of drug paraphernalia
Fraud, forgery, and counterfeiting	Deceptive practices or identification, fraud, giving false information, altering or forging documents, and counterfeiting or possession of counterfeit materials or tools
Homicide	Murder, manslaughter, and homicide
Immigration	Illegal entry, illegal reentry, false claim to U.S. citizenship, alien smuggling, and removal proceedings; offenses include both criminal offenses and civil immigration violations
Kidnapping	False imprisonment, kidnapping, and taking hostages
Larceny/theft	Grand and petty larceny and theft, shoplifting, embezzlement, and money laundering
Motor vehicle theft	Auto theft, carjacking, and taking a vehicle without consent
Obstruction of justice	Escaping, evading, being a fugitive of justice, failing to appear, failing to register as a sex offender, resisting arrest, and interfering with or obstructing an officer or justice proceedings
Property damage	Destruction of property, vandalism, and criminal or malicious mischief
Robbery	Armed robbery, robbery of a dwelling, robbery of a bank, and unarmed robbery
Sex offenses	Lewd and lascivious acts, rape, sexual assault, indecent exposure, prostitution, and molestation
Stolen Property	Buying, selling, receiving, or possessing stolen property
Terrorism	Terrorism-related offenses
Traffic violations	Driving under the influence, hit and run, no proof of insurance, no driver's license, and moving violations such as speeding and failure to stop
Weapons violations	Possession of a weapon, discharging a weapon, altering a weapon, and carrying a concealed weapon
Miscellaneous	Any other offense not listed above, which may include trespassing, gang participation, child cruelty, and alcohol-related offenses, as well as offenses for which not enough information were available (i.e., "arrest by a law enforcement officer")

Source: GAO. | GAO-18-433.

Note: Offenses include an attempt or conspiracy to commit the respective offense. We developed the criminal offense categories based on our prior work that used the FBI's classification for offense codes as our guidance. See GAO-11-187.

Table 6. Bureau of Prisons Primary Offense Categories, Attempted and Committed

Primary offense category	Category includes
Burglary/larceny	Burglary; violations of certain customs laws; racketeering; theft; theft of government property and funds; motor vehicle theft; robbery; Interstate Commerce Act violations; stalking; theft involving intellectual property
Continuing criminal enterprise	Continuing criminal enterprise
Counterfeit/embezzlement	Counterfeiting money, postage, and similar offenses; embezzlement; violations of anti-trust laws; violations of certain elections laws; violations of certain banking and insurances laws
Court/corrections	Escaping; bail/bond jumping; criminal contempt; mutiny/riot in a federal penal facility; contraband in prison; perjury; obstructing justice
Drugs	Simple possession of a controlled substance; manufacturing, distributing, or possessing with intent to manufacture or distribute a controlled substance; importing narcotics; obtaining a controlled substance by fraud; distribution to a minor; investment of illicit drug profits; maintaining drug-involved premises
Fraud/bribery/extortion	Bribery; extortion; forgery; various types of fraud offenses, including tax fraud
Homicide/aggravated assault	Homicide; death or injury caused by explosives; assault; assaulting a federal officer; kidnapping; threats against the President; terrorism; domestic violence; drive-by shooting; violence against maritime navigation and similar offenses; criminal street gangs
Immigration	Illegal reentry; illegal entry; alien smuggling; impersonation; passport fraud
National security	War crimes; espionage; sedition; treason; biological weapons; desertion; impersonating foreign diplomats, consuls, or officers
Robbery	Robbery offenses, including bank robbery
Sex offenses	Child sexual abuse; mailing obscene matter; sexual abuse; rape; prostitution; incest; failure to register as a sex offender; sexual abuse resulting in death; possession of child pornography
Weapons/explosives	Carrying a deadly weapon; arson; participation in nuclear and weapons of mass destruction threats to the United States; engaging in the business of importing, manufacturing or dealing in explosives without a license and related offenses; firearms offenses
Miscellaneous	Discharging pollution; destruction of energy facilities; liquor law violations; disorderly conduct; vagrancy; damage to religious property and similar offenses; violations of food and drug laws; violations of fishing and gaming laws

Source: GAO analysis of Bureau of Prisons data. | GAO-18-433.

Note: Offenses include an attempt or conspiracy to commit the respective offense.

In addition, we analyzed aggregated U.S. Sentencing Commission data on the approximately 201,300 federal convictions of criminal aliens from fiscal years 2011 through 2016.[98] These criminal aliens may have also been included in our federal study population if they were convicted in federal court and sentenced in fiscal years 2011 through 2016 and also incarcerated between these same fiscal years. Table 7 describes the primary offense categories for federal convictions used by the U.S. Sentencing Commission, which we combined into 15 larger categories for our purposes.[99]

In addition, to determine the types of primary offenses for which SCAAP criminal aliens were convicted, we analyzed conviction data from five state prison systems—Arizona, California, Florida, New York, and Texas—for the most recent fiscal year in which data were available (2015). We selected these five state prison systems because they had the most SCAAP criminal alien incarcerations in state prisons in fiscal year 2015. Collectively, these five state prison systems accounted for 64 percent of the SCAAP criminal alien incarcerations in state prisons during fiscal year

[98] The U.S. Sentencing Commission data are limited to felony and Class A misdemeanor cases for offenders who are convicted and sentenced in the federal court system. These data do not include state cases, federal petty offenses, federal cases which result in all charges being dismissed or acquitted, federal death penalty cases, federal juvenile cases, or federal witness protection cases. They also do not include convicted offenders for whom no sentences were yet issued, offenders sentenced but for whom no sentencing documents were submitted to the Commission, and offenders sentenced prior to the enactment of the Sentencing Reform Act of 1984. See Pub. L. No. 98-473, ch. II, 98 Stat. 1837, 1987-2001 (1984). According to U.S. Sentencing Commission officials, data do include a small number of individuals that may have been fined or given probation instead of a federal prison sentence. Convictions that did not have information on offender's citizenship status were excluded. From fiscal years 2011 through 2016, 2 percent of the nearly 456,000 convictions were excluded for this reason. Information on the citizenship status of offenders is obtained from their presentence investigation report.

[99] For our reporting purposes, we consolidated certain U.S. Sentencing Commission categories into the following categories: Drugs includes drugs - trafficking, drugs - communication facility, and drugs - simple possession; Economic crimes includes larceny, fraud, embezzlement, forgery/counterfeiting, tax, and antitrust; Sex offenses includes sexual abuse and child pornography; Money laundering/ racketeering/ extortion includes money laundering and racketeering/extortion; Homicide includes murder and manslaughter; and the "other" category includes bribery, gambling/lottery, civil rights, prison offenses, environmental/wildlife, national defense, food and drug, and other miscellaneous offenses.

2015.[100] They are also the same prison systems that we analyzed in our 2011 report on criminal aliens.[101] The information obtained from the selected state prison systems is not generalizable to all state prison systems, but provides useful insights about why SCAAP criminal aliens were incarcerated.[102] For these state prison systems, we analyzed the conviction data for the SCAAP criminal aliens incarcerated in fiscal year 2015.[103] Table 8 describes the primary offense categories for state convictions.[104]

Table 7. Primary Offense Categories, as Reported by the U.S. Sentencing Commission

Primary offense category	Category includes
Administration of Justice offenses	Commission of offense while on release, bribery of a witness, failure to appear by offender, contempt, failure to appear by material witness, obstruction of justice, payment of witness, perjury or subornation of perjury, misprision of a felony, and accessory after the fact
Antitrust	Bid-rigging, price-fixing, and market allocation agreement
Arson	Arson, including damage by explosives
Assault	Attempt to commit murder, assault with intent to murder, threatening communication, aggravated assault, conspiracy with attempt to murder, obstructing or impeding officers, minor assault, and conspiracy that includes assault with attempt to murder
Auto theft	Auto theft (including parts), receipt/possession of stolen auto or parts, and altered identification numbers/trafficking in altered (auto)
Bribery	Payment to obtain office, bribe involving officials, bribery - bank loan/commercial, loan or gratuity to bank examiner, etc., gratuity involving officials, and bribe or gratuity affecting employee plan
Burglary/breaking & entering	Post office burglary, burglary of [Drug Enforcement Administration] DEA premises (pharmacy), burglary of other structure, bank burglary, and burglary of a residence

[100] Our analysis included state prison systems that participated in SCAAP and did not include U.S. territories. State prison systems in Arkansas, West Virginia, Vermont, and the District of Columbia did not receive reimbursement for SCAAP criminal aliens incarcerated in fiscal year 2015.

[101] GAO-11-187.

[102] We did not examine the extent to which state policies may have affected the number of individuals convicted of crimes in each state. Further, because policies may vary by state, we presented data on each state separately.

[103] Each state's definition of a primary offense for which an individual could be convicted is detailed later in this report.

[104] Each state categorized the primary offenses according to the categories, except for New York. We categorized New York's offenses and obtained their concurrence on our categorization.

Table 7. (Continued)

Primary offense category	Category includes
Child pornography	The sale, distribution, transportation, shipment, receipt, or possession of materials involving the sexual exploitation of minors
Civil rights	Interference with rights under color of law; force or threats to deny benefits or rights; obstructing an election or registration; manufacture, etc. - eavesdropping device; other deprivations/discrimination; obstructing correspondence; peonage, servitude, and slave trade; intercept communication or eavesdropping; and conspiracy to deprive individual of civil rights
Drugs—communication facility	Use of a communication facility in a drug trafficking offense
Drugs—simple possession	Distribution of a small amount of marijuana and simple possession
Drugs—trafficking	Drug distribution/manufacture; drug distribution/manufacture - conspiracy, continuing criminal enterprise, drug distribution; employee under 21, drug distribution near school; drug import/export; drug distribution to person under 21;establish/rent drug operation, and distributing, importing; exporting listed chemicals
Environmental/wildlife	Waste discharge, specially protected fish, wildlife, and plants
Embezzlement	Embezzlement - property, embezzlement from labor unions, embezzlement - mail/post office, embezzlement from benefit plans, and bank embezzlement
Firearms	Unlawful possession/transportation of firearms or ammunition; possession of guns/explosives on aircraft; unlawful trafficking, etc., in explosives; possession of guns/explosives in federal facility/schools; use of fire or explosives to commit felony; and use of firearms or ammunition during crime
Food and drug	False information or tampering with products, tampering to injure business, tampering with risk of death or injury, and violation of regulations involving food, drugs, etc.
Fraud	Odometer laws and regulations, insider trading, and fraud and deceit
Forgery/counterfeiting	Counterfeit bearer obligations and forgery/counterfeit (non-bearer obligations)
Gambling/lottery	Engaging in a gambling business, transmission of wagering information, obstruction to facilitate gambling, and interstate transportation of wagering paraphernalia
Immigration	Trafficking in U.S. passports; trafficking in entry documents; failure to surrender naturalization certificate; fraudulently acquiring U.S. passports; smuggling, etc.; unlawful alien; fraudulently acquiring entry documents; and unlawfully entering the United States
Kidnapping/hostage taking	Ransom taking and hostage/kidnapping
Larceny	Bank larceny, theft from benefit plans, other theft - mail/post office, receipt/possession of stolen property (not auto), other theft - property, larceny/theft/mail/post office, larceny/theft property (not auto), and theft from labor union
Manslaughter	Involuntary and voluntary manslaughter

Primary offense category	Category includes
Money laundering	Laundering of monetary instruments, monetary transaction from unlawful activity, failure to file currency report, and failure to report monetary transactions
Murder	First degree murder, felony with death resulting, second degree murder, conspiracy to murder (with death resulting)
National defense	Evasion of export controls and exportation of arms, etc., without license
Prison offenses	Contraband in prisons, riots in federal facilities, and escape
Racketeering/extortion	Extortionate extension of credit, blackmail, extortion by force or threat, Hobbs Act extortion, travel in aid of racketeering, crime relating to racketeering, and violent crimes in aid of racketeering
Robbery	Bank robbery, aggravated bank robbery, Hobbs Act robbery, mail robbery, other robbery, and carjacking
Sexual abuse	Sexual abuse of a minor; transportation of minor for sex; sexual abuse of a ward; criminal sexual abuse; abusive sexual contact
Tax	Receipt/trafficking in smuggled property, aid, etc., in tax fraud; fraud - tax returns, statements, etc.; fraud, false statement - perjury; failure to file or pay; tax evasion; evading import duties (smuggling); failure to collect or account for taxes; regulatory offenses - taxes; failure to deposit taxes in trust account; non-payment of taxes; conspiracy to avoid taxes; and offenses relating to withholding statements
Other miscellaneous offenses	Illegal use of regulatory number - drugs; illegal transfer of drugs; illegal regulatory number to get drugs; drug paraphernalia; forgery/fraud for drugs; dangerous devices to protect drugs; manufacture drugs against quota; endangering life while manufacturing drugs; operate carrier under drugs; endangerment from hazardous/toxic substances; mishandling substances, records, etc.; threat of tampering with public water system; hazardous devices on federal lands; mishandling other pollutants, records, etc.; improper storage of explosives; recordkeeping violation - explosives; possession of other weapon - on aircraft, in federal facility; failure to report theft of explosives; feloniously mailing injurious articles; transport of hazardous material in
	commerce; interference with flight crew, other offense - aboard aircraft; criminal infringement of copyright/trademark; conflict of interest; unauthorized payment; non-drug forfeiture; impersonation; false statement to Employee Act; reporting offenses - labor related; criminal infringement of trademark; unlawful conduct relating to control/cigarettes; trespass; destruction of property; destruction of mail; aircraft piracy; conspiracy to murder (no death, assault, or attempt); conspiracy to commit murder; and all other miscellaneous offenses not previously listed in any of the other categories

Source: U.S. Sentencing Commission. | GAO-18-433.

Note: The U.S. Sentencing Commission categorizes offenses into the major categories above. For our reporting purposes, we consolidated certain U.S. Sentencing Commission categories into the following categories: Drugs includes drugs - trafficking, drugs - communication facility, and drugs - simple possession; Economic crimes includes larceny, fraud, embezzlement, forgery/counterfeiting, tax, and antitrust; Sex offenses includes sexual abuse and child pornography; Money laundering/racketeering/extortion includes money laundering and racketeering/extortion; Homicide includes murder and manslaughter; and the "other" category includes bribery, gambling/lottery, civil rights, prison offenses, environmental/wildlife, national defense, food and drug, and other miscellaneous offenses.

- Finally, to determine what is known about certain individuals with international terrorism-related convictions, we analyzed Department of Homeland Security (DHS), DOJ, and U.S. Sentencing Commission data on individuals on the DOJ's National Security Division (NSD) list of individuals with public and unsealed federal convictions resulting from international terrorism-related investigations from March 19, 2010 through December 31, 2016.[105] We report on this analysis in appendix V. We collected data on alienage or citizenship status, country of birth, naturalization date, and extradition status, as available, from BOP, U.S. Sentencing Commission, U.S. Citizenship and Immigration Services (USCIS), and ICE, using identifiers that we matched across agencies.[106] For example, USCIS records would not typically contain information about the citizenship status of individuals who are U.S. citizens and were born in the United States, but would contain information about naturalized citizens or aliens who had applied for a benefit with USCIS. We considered the totality of data available from BOP's SENTRY database, the U.S. Sentencing Commission, USCIS's Central Index System, and ICE's Enforcement Integrated Database to determine (a) the alienage or citizenship status of individuals at the time of their conviction and (b) the country of birth for each individual on the NSD list with a conviction from March 19, 2010 through 2016 for whom data was available.[107] Specifically: We identified those

[105] Specifically, we analyzed all convictions resulting from international terrorism investigations after March 18, 2010—the last date we analyzed for our 2011 report— through December 31, 2016—the most recent data available for our audit.

[106] The NSD list includes names of individuals convicted as a result of international terrorism-related investigations and does not include additional identifiers for these individuals. We obtained and matched identifiers iteratively across BOP, U.S. Sentencing Commission, USCIS, and ICE data using the names on the NSD list as a starting point. The identifiers that we matched on included: name, date of birth, country of birth, FBI number, and/or alien registration number.

[107] According to USCIS, the Central Index System contains information on the status of applicants/petitioners seeking immigration benefits, to include: lawful permanent residents, naturalized citizens, U.S. border crossers, aliens who have been issued employment authorization documents, individuals who petitioned for benefits on behalf of family members, and other individuals subject to the provisions of the Immigration and Nationality

individuals that the U.S. Sentencing Commission indicated were aliens that had been extradited to the United States for prosecution as extradited aliens.

- We identified those individuals for whom USCIS provided a date of naturalization as naturalized citizens, and compared their date of naturalization with the date that they were charged with an international terrorism-related crime to determine if they naturalized before being charged. We used USCIS data to identify the country of birth for these individuals.
- We determined that those other individuals for whom USCIS and ICE had a record that indicated they were citizens of a country other than the United States were aliens and used USCIS and U.S. Sentencing Commission data to identify their country of birth.
- We reviewed the data for the remaining individuals and determined that those individuals U.S. Sentencing Commission and/or BOP identified as U.S.-born or as U.S. citizens were U.S. citizens.
- We identified individuals for whom we could not determine a country of birth or country of citizenship using the totality of information available as unknown.

According to DOJ, the list includes both individuals convicted of crimes that DOJ considers to be directly related to international terrorism and individuals convicted of a variety of other crimes where the investigation, at the time of charging, appeared to involve an identified link to international terrorism. We did not independently verify these individuals' links to terrorism.

Act. According to ICE, the Enforcement Integrated Database maintains information related to the investigation, arrest, booking, detention, and removal of persons encountered during immigration and criminal law enforcement investigations and operations conducted by certain Department of Homeland Security components, namely ICE and U.S. Customs and Border Protection.

Table 8. State Primary Offense Categories, Attempted and Committed

Primary offense category	Category includes
Arson	Arson, reckless burning, and possession of arson materials
Assault	Assault, battery, assault with a deadly weapon, endangerment, and threats
Burglary	Breaking and entering, burglary, and possession of burglary tools
Disorderly Conduct	Disturbing the peace, fighting, intoxication, public nuisance, and disorderly conduct
Drugs	Use/under the influence, possession, possession with intent to distribute, sales, manufacturing, transporting, and possession of drug paraphernalia
Fraud, forgery, and counterfeiting	Deceptive practices or identification, fraud, giving false information, altering or forging documents, and counterfeiting or possession of counterfeit materials or tools
Homicide	Murder, manslaughter, and homicide
Immigration	Illegal entry, illegal reentry, false claim to U.S. citizenship, alien smuggling, and removal proceedings
Kidnapping	False imprisonment, kidnapping, and taking hostages
Larceny/theft	Grand and petty larceny and theft, shoplifting, embezzlement, and money laundering
Motor vehicle theft	Auto theft, carjacking, and taking a vehicle without consent
Obstruction of justice	Escaping, evading, being a fugitive of justice, failing to appear, failing to register as a sex offender, resisting arrest, and interfering with or obstructing an officer or justice proceedings
Property damage	Destruction of property, vandalism, and criminal or malicious mischief
Robbery	Armed robbery, robbery of a dwelling, robbery of a bank, and unarmed robbery
Sex offenses	Lewd and lascivious acts, rape, sexual assault, indecent exposure, prostitution, and molestation
Terrorism	Terrorism-related offenses
Traffic violations	Driving under the influence, hit and run, no proof of insurance, no driver's license, and moving violations such as speeding and failure to stop
Stolen Property	Buying, selling, receiving, or possessing stolen property
Traffic violations	Driving under the influence, hit and run, no proof of insurance, no driver's license, and moving violations such as speeding and failure to stop
Weapons violations	Possession of a weapon, discharging a weapon, altering a weapon, and carrying a concealed weapon
Miscellaneous	Any other offense not listed above, as determined by each state

Source: GAO. | GAO-18-433.

Note: Offenses include an attempt or conspiracy to commit the respective offense. We developed the criminal offense categories based on our prior work that used the FBI's classification for offense codes as our guidance.

Criminal Alien Incarceration Costs

To determine the costs associated with incarcerating the criminal aliens in the United States, we obtained and analyzed cost and inmate data from BOP, data on SCAAP reimbursements for incarcerating SCAAP criminal aliens, and cost and inmate data from states and localities. Specifically, we analyzed costs to (1) the federal government, (2) state prison systems, and (3) selected states and localities. We calculated the total annual cost to the federal government, for fiscal years 2010 through 2015, to incarcerate criminal aliens, by combining: (1) the estimated costs to incarcerate criminal aliens in federal prisons and (2) the costs to reimburse states and localities for incarcerating SCAAP criminal aliens. Specifically, we used the average of the BOP's 12 monthly snapshot population data from SENTRY for each type of BOP institution for each fiscal year, and multiplied these population data by BOP per capita costs for each facility type to determine the total federal cost to incarcerate criminal aliens in federal prisons. In addition, we used SCAAP reimbursement data for SCAAP fiscal years 2011 through 2016 to determine the cost to the federal government to reimburse states and localities for incarcerating SCAAP criminal aliens in state prisons and local jails.[108]

For all 50 state prison systems, we estimated selected operating costs associated with incarcerating SCAAP criminal aliens from fiscal years 2010 through 2015 using Bureau of Justice Statistics and SCAAP reimbursement data as well as inflation factors.[109] These selected operating costs include correctional officer salaries, medical care, food service, and utilities. The salaries for correctional officers were obtained for each year from SCAAP data. The other three categories were calculated using

[108] Each year's SCAAP program is based on SCAAP criminal aliens incarcerated during the prior fiscal year, July 1 through June 30. For example, the fiscal year 2016 SCAAP program will include SCAAP criminal aliens that were incarcerated in state prisons and local jails between July 1, 2014 and June 30, 2015.

[109] Bureau of Justice Statistics data was obtained from a Bureau of Justice Statistics study that estimated state prison expenditures for medical care, food service, and utilities for all 50 state prison systems in 2001. Bureau of Justice Statistics, *State Prison Expenditures*, 2001 (Washington, D.C.: June 2004). These were the most recent cost data available for our purposes. Data were not available on local jail systems.

Bureau of Justice Statistics' 2001 state prison expenditure data, the total number of inmate days in a given year by state as reported through SCAAP, and annual IHS Global Insight price deflators for the corresponding categories.[110] We multiplied the estimated selected operating costs per inmate by the number of SCAAP inmate days to determine the total costs for incarcerating SCAAP criminal aliens for each state prison system. While our estimates provide insight into state expenditures to incarcerate SCAAP criminal aliens, our estimates may not represent actual costs if per capita prison expenditures for incarcerating criminal aliens grew at a different rate than the inflation factors that we used for each category. In addition to using estimated operating costs for medical care, food service, and utilities, we also used data on correctional officer salaries from SCAAP. Further, our estimation may not include all related incarceration costs and therefore may not reflect actual costs. For example, using actual average daily cost data for a selected set of state prison systems, we found that in 2015 costs using these average daily costs were higher than the estimated costs using inflation factors.[111] Our analysis includes those state prison systems that received SCAAP reimbursement in the associated fiscal year. As a result, not all 50 state prison systems may be included in each fiscal year.[112] Our analysis also includes the District of Columbia. Moreover, we used the GDP deflator to present criminal alien costs in fiscal year 2016 dollars.

To estimate the costs associated with incarcerating SCAAP criminal aliens in selected state prison and local jail systems, we obtained and analyzed cost and inmate data from five state prison systems (Arizona, California, Florida, New York, and Texas) and five local jail systems (Maricopa County, Arizona; Orange County, California; Los Angeles

---

[110] IHS Global Insight is a firm that provides comprehensive economic and financial information on countries, regions, and industries.

[111] In addition to the costs for incarcerating criminal aliens, there are other costs associated with criminal aliens, such as the costs necessary to identify, apprehend, detain, process through immigration court, and remove an individual from the country.

[112] The following state prison systems did not receive SCAAP reimbursement: West Virginia in fiscal year 2010; West Virginia and the District of Columbia in fiscal years 2011 and 2012; Missouri, Tennessee, Wyoming, West Virginia, and the District of Columbia in fiscal year 2013; West Virginia and the District of Columbia in fiscal year 2014; and Arkansas, West Virginia, Vermont, and the District of Columbia in fiscal year 2015.

County, California; Essex County, New Jersey; and Harris County, Texas)—those with the highest number of SCAAP criminal alien incarcerations. Specifically, we estimated the amount that each of these state prison and local jail systems expended to incarcerate SCAAP criminal aliens based on average daily cost data and the number of SCAAP criminal alien inmate days.[113] For fiscal year 2015 these five state prison systems and the six local jail systems accounted for 64 percent and 19 percent of SCAAP criminal alien incarcerations in state prisons and local jails, respectively. However, we did not include cost estimates for one of the top six local jail systems, New York City, New York.

Officials from this locality stated that they no longer apply for SCAAP funds, and they did not provide us an average daily cost per inmate. These state prison and local jail systems are the same, with the exception of Essex County, New Jersey, as those we selected in our April 2005 and March 2011 reports.[114] While our analysis provides insights into the costs associated with incarcerating SCAAP criminal aliens in these state prison and local jail systems and we are able to use the actual average daily costs provided by each entity, along with actual SCAAP criminal alien prison population data, the results of this analysis are not generalizable to other state prison and local jail systems. Also, given that these estimated costs are based on average daily cost for the general prison population, they may not represent actual costs if expenditures on the criminal alien prison population differ from the average daily costs; moreover, state prison and local jail systems might not include all prison expenditures in the provided average daily costs and what is included might differ across state prison and local systems.

---

[113] Since, the costs from state prison and local jail systems correspond to each jurisdiction's unique fiscal year, which is different than the federal fiscal year (October 1 through September 30) we used a weighted average cost. Therefore, state and local fiscal year costs were weighted by the number of months that overlapped with the federal fiscal year.

[114] GAO-11-187 and GAO, *Information on Criminal Aliens Incarcerated in Federal and State Prisons and Local Jails*, GAO-05-337R (Washington, D.C.: Apr. 7, 2005).

Criminal Alien Removability from the United States and Experiences of Aliens after Incarceration in Federal Prison

To determine what is known about the removability from the United States of criminal aliens incarcerated in federal prison and the experiences of criminal aliens after their incarceration in federal prison, we matched data from criminal aliens incarcerated in federal prison with other available data sources using various identifiers, including alien numbers and FBI numbers. We reviewed the quality of the matches using date of birth and/or name data, where available. We also reviewed ICE's Enforcement and Removal Operations reports from fiscal years 2015 and 2016, which included information about the number of ICE removals of aliens in the United States from fiscal years 2011 through 2016, and the proportion of such removals that were criminal aliens compared to noncriminal aliens. We report on this analysis in appendix IX.

Specifically, to analyze the potential removability from the United States of criminal aliens at the time they were incarcerated in federal prison, we matched BOP data on criminal aliens incarcerated from fiscal years 2011 through 2016 with data from ICE's Alien Criminal Response Information Management System (ACRIMe). ICE specialists use ACRIMe to provide an indication of an individual's identity and potential removability to law enforcement partners, at the request of the law enforcement partner. We used FBI numbers and/or alien numbers as unique identifiers to match these data. We then verified that the matches appeared reasonable by comparing names and dates of birth of those individuals who matched using the unique identifiers. To determine each criminal alien's potential removability from the United States at the time they entered federal prison, we identified the ACRIMe record for each criminal alien, as available, that was created closest to the date the criminal alien first entered federal prison and used this record for our analysis. We restricted our analysis to those ACRIMe records that were generated within 60 days of the date the criminal alien entered federal prison because a criminal alien's removability from the United States can change over time, including while the alien is incarcerated.

To determine what is known about the experience of criminal aliens after their incarceration in federal prison, we matched data from those criminal aliens who completed a term of incarceration in federal prison from fiscal years 2011 through 2016 with DHS and DOJ data. We used this information to determine if these criminal aliens were subsequently removed from the United States, reincarcerated, and/or received naturalized citizenship. We compared criminal aliens' federal incarceration completion date with the dates of federal government and/or law enforcement agency encounters to determine if those encounters took place after the alien's completion of their term of incarceration in federal prison. Specifically, we matched FBI and/or alien numbers, as available, of criminal aliens who completed a term of incarceration in federal prison from fiscal years 2011 through 2016 with:

- ICE Enforcement Integrated Database data from fiscal year 2011 through June 2017 to determine if the alien was removed by ICE after completion of their term of incarceration in federal prison;[115]
- BOP SENTRY data from fiscal years 2011 through 2016 to determine if a criminal alien was reincarcerated in federal prison after completion of their term of incarceration in federal prison;
- BJA SCAAP data from fiscal years 2010 through 2015, to determine if a criminal alien was subsequently incarcerated in a state prison or local jail system participating in SCAAP after completion of their term of incarceration in federal prison; and
- USCIS Central Index System data from fiscal year 2011 through July 2017 to determine if a criminal alien received naturalized citizenship after completion of their term of incarceration in federal prison.

---

[115] ICE provided the most recent removal date in its records for each criminal alien, as of June 2017. Some aliens may have been removed multiple times between fiscal year 2011 and June 2017. An alien removed multiple times would only appear in the data with their most recent removal date.

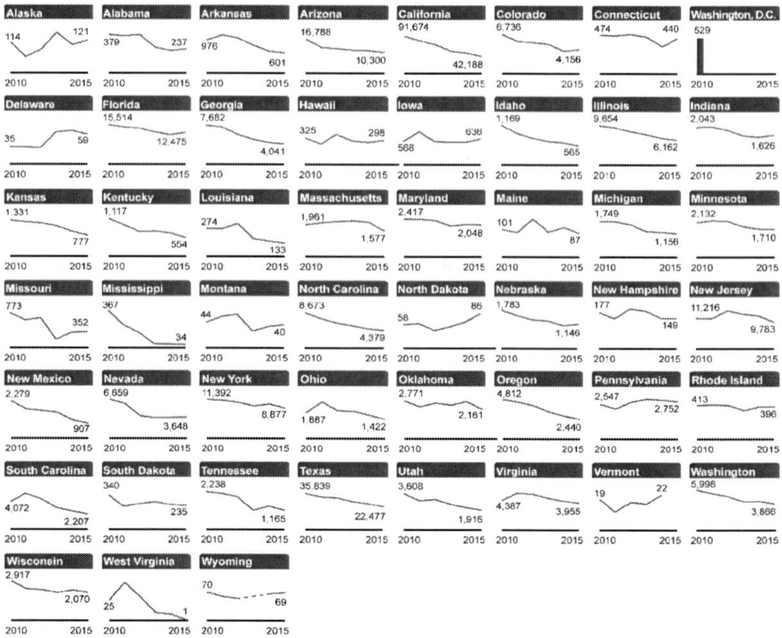

Source: GAO analysis of Bureau of Justice Assistance data. | GAO-18-433.

Note: The above figure includes SCAAP criminal alien incarcerations in state prisons and local jails in each of the 50 states and District of Columbia. Because of differences in the number of incarcerations by state, each state may be on a different scale. The following states did not have a state or locality that applied for SCAAP in certain fiscal years: District of Columbia (2011 through 2015), Vermont (2015), and Wyoming (2013). In addition, U.S. territories that received SCAAP reimbursements are not included in the above figure. SCAAP data represent the number of incarcerations, rather than number of SCAAP criminal aliens since these aliens could have multiple SCAAP incarcerations in the same fiscal year. As stated previously, SCAAP criminal aliens represent a portion of the total population of criminal aliens incarcerated in state prisons and local jails.

Figure 16. Number of State Criminal Alien Assistance Program (SCAAP) Criminal Alien Incarcerations by State from Fiscal Years 2010 through 2015.

We determined that the data used in each of our analyses were sufficiently reliable for the purposes of this chapter by analyzing available documentation, such as related data dictionaries; interviewing officials knowledgeable about the data; conducting electronic tests to identify missing data and anomalies; and following up with officials, as appropriate.

We conducted this performance audit from August 2016 to July 2018 in accordance with generally accepted government auditing standards. Those standards require that we plan and perform the audit to obtain sufficient, appropriate evidence to provide a reasonable basis for our findings and conclusions based on our audit objectives. We believe that the evidence obtained provides a reasonable basis for our findings and conclusions based on our audit objectives.

Appendix II: State Criminal Alien Assistance Program Criminal Alien Incarcerations in State Prisons and Local Jails

Figure 16 shows the number of State Criminal Alien Assistance Program (SCAAP) criminal alien incarcerations in state prisons and local jails in each of the 50 states and District of Columbia from fiscal years 2010 through 2015. Criminal alien incarcerations are included in the figure below if the state or locality received SCAAP reimbursement. Therefore, these data represent only a portion of all criminal alien incarcerations in state prisons and local jails.

Appendix III: Arrests/Transfers of Criminal Aliens by Type of Arresting Agency

This appendix provides additional details on arrests/transfers of criminal aliens in our federal study population and State Criminal Alien Assistance Program (SCAAP) criminal aliens in our state and local study population. The data did not allow us to determine the difference between a new arrest and a transfer from one arresting agency to another; as a result, we refer to these collectively as "arrests/transfers." To determine the arrest history of criminal aliens incarcerated in federal and state prisons and local jails, we pulled a random sample of 500 criminal aliens from our federal study population and 500 SCAAP criminal aliens from our state and local study population, and, using data from the Federal Bureau of Investigation (FBI), estimated arrest data for each of our study populations.[116] Specifically, we break down the estimated number of arrests/transfer and estimated number of arrest offenses by those made by (1) federal arresting agencies and (2) state and local arresting agencies. While the samples we selected for our analyses allowed us to estimate and provide valuable insights about the arrest history of the approximately 197,000 criminal aliens in our federal study population and the approximately 533,000 SCAAP criminal aliens in our state and local study population, our analyses are not generalizable to the arrest history of criminal aliens not in these study populations.

Federal Criminal Alien Population

Based on our random sample of criminal aliens incarcerated in federal prisons during fiscal years 2011 through 2016, we estimated that our federal study population of about 197,000 criminal aliens accounted for about 761,300 arrests/transfers by federal arresting agencies. This same population accounted for about 589,800 arrests/transfers state or local arresting agencies from 1974 through 2017.[117] We estimated that 91

---

[116] For these analyses, we only included criminal aliens in the study population if they had an FBI number available, as this was needed to match data across databases. As a result, the study populations that we projected to for these analyses were smaller than the originating study populations. For example, our federal study population started with approximately 198,000 criminal aliens and about 197,000 had FBI numbers. Therefore, for this analysis, we included approximately 197,000 criminal aliens in our federal study population.

[117] These dates, 1974 through 2017, represent the dates of the oldest and newest records in our FBI dataset. Law enforcement entities send arrest information to the FBI on a voluntary basis; as a result, FBI data on arrest history may not include all arrests. Because our sample populations are separate for our federal criminal alien population and SCAAP criminal alien population, results cannot be compared to the results we presented in our 2011 report. See GAO, *Criminal Alien Statistics: Information on Incarcerations, Arrests, and Costs,* GAO-11-187 (Washington, D.C.: Mar. 24, 2011). All percentage estimates presented in this report have a margin of error of plus or minus 7 percentage points or fewer. All estimates of the

percent of the arrests/transfers by federal arresting agencies and 67 percent of the arrests/transfers by state and local arresting agencies occurred from 2000 through 2017.

We estimated that the approximately 197,000 criminal aliens in our federal study population were arrested/transferred by federal arresting agencies for a total of about 1 million offenses.[118] A single arrest can be for multiple offenses, and being arrested for one or more offenses does not necessarily result in prosecution or a conviction of all, or any, of the offenses for which an individual was arrested. Of the approximately 1 million offenses, we estimated that 81 percent were related to immigration, as shown below in Table 9.[119] Each offense category in the table may include an attempt or conspiracy to commit the respective offense.

We estimated that the approximately 197,000 criminal aliens in our federal study population were arrested/transferred by state and local arresting agencies for a total of about 996,700 offenses. Of these offenses, we estimated that 41 percent were related to drugs or traffic violations, as shown above in Table 10. Each offense category in the table may include an attempt or conspiracy to commit the respective offense.

---

number of arrests/transfers or offenses have a relative error of plus or minus 14 percent of the estimate or less. See appendix I for more details on the margin of error for these estimates.

[118] For our analysis, (a) multiple counts of the same offense were counted once, (b) duplicate offenses on the same day may be counted more than once if the person was transferred to another agency on the same day and that agency submitted the same offense to the FBI, and (c) duplicate offenses on the same day submitted by the same agency may be counted more than once if there was not enough information to determine that two offenses were the same. Because we selected samples from both our federal study population and our state and local study population, results cannot be compared to the results we presented in our 2011 report, see GAO-11-187.

[119] See appendix I for a complete description of each of the offense categories.

Table 9. Estimated Number and Percent of Attempted or Committed Offenses for Which Criminal Aliens Incarcerated in Federal Prisons from Fiscal Years 2011 through 2016 who had an FBI Number Were Arrested/Transferred by Federal Arresting Agencies from 1974 through 2017

Arrest offense	Estimated number	Estimated percent
Immigration[a]	863,700	81.0
Drugs	133,400	12.5
Obstruction of justice	25,400	2.4
Fraud, forgery, and counterfeiting	12,300	1.2
Weapons violations	7,500	0.7
Miscellaneous	6,400	0.6
Larceny/theft	4,000	0.4
Traffic violations	4,000	0.4
Assault	3,200	0.3
Robbery	2,000	0.2
Kidnapping	1,200	0.1
Motor vehicle theft	1,200	0.1
Sex offenses	1,200	0.1
Burglary	800	0.1
Property damage	800	0.1
Terrorism	400	< 0.1
Arson	--	--
Disorderly Conduct	--	--
Homicide	--	--
Stolen Property	--	--
Total[b]	1,067,400	100

Legend:-- = no estimated offenses.

Source: GAO analysis of Federal Bureau of Investigation (FBI) data. | GAO-18-433.

Note: Offenses may include an attempt or conspiracy to commit the respective offense. While our analyses allowed us to estimate and provide valuable insights about the arrest history of our study population, which consisted of about 197,000 criminal aliens incarcerated in federal prisons from fiscal years 2011 through 2016 with an FBI number, our analyses are not intended to infer conclusions about the arrest history of criminal aliens not in this study population. Law enforcement entities send arrest information to the FBI on a voluntary basis; as a result, FBI data on arrest history may not include all arrests. The numbers in the table above represent the number of offenses we estimated that criminal aliens were arrested or transferred for; they do not represent the number of times that criminal aliens were arrested or transferred for each offense. All estimates in this table have a margin of error of +/- 3 percentage points or fewer.

[a]Offenses included in our immigration category include both criminal immigration offenses and civil immigration violations— administrative grounds of removability.

[b]Numbers may not sum to totals because of rounding. Percentages may not sum to 100 due to rounding.

Table 10. Estimated Number and Percent of Attempted or Committed Offenses for Which Criminal Aliens Incarcerated in Federal Prisons from Fiscal Years 2011 through 2016 who had an FBI Number Were Arrested/Transferred by State and Local Arresting Agencies from 1979 through 2017

Arrest offense	Estimated number	Estimated percent
Drugs	203,200	20.4
Traffic violations	200,500	20.1
Obstruction of justice	115,900	11.6
Assault	105,200	10.6
Miscellaneous	67,900	6.8
Larceny/theft	66,300	6.7
Fraud, forgery, and counterfeiting	50,000	5.0
Burglary	44,100	4.4
Weapons violations	37,000	3.7
Motor vehicle theft	18,300	1.8
Property damage	16,700	1.7
Stolen property	14,300	1.4
Disorderly Conduct	12,300	1.2
Sex offenses	12,300	1.2
Robbery	11,500	1.2
Immigration[a]	10,700	1.1
Homicide	6,000	0.6
Kidnapping	4,400	< 0.1
Arson	400	< 0.1
Terrorism	--	--
Total[b]	996,700	100

Legend: -- = no estimated offenses.

Source: GAO analysis of Federal Bureau of Investigation (FBI) data. | GAO-18-433.

Note: Offenses may include an attempt or conspiracy to commit the respective offense. While our analyses allowed us to estimate and provide valuable insights about the arrest history of our study population, which consisted of about 197,000 criminal aliens incarcerated in federal prisons from fiscal years 2011 through 2016 with an FBI number, our analyses are not intended to infer conclusions about the arrest history of criminal aliens not in this study population. Law enforcement entities send arrest information to the FBI on a voluntary basis; as a result, FBI data on arrest history may not include all arrests. The numbers in the table above represent the number of offenses we estimated that criminal aliens were arrested or transferred for; they do not represent the number of times that criminal aliens were arrested or transferred for each offense. All estimates in this table have a margin of error of +/- 4 percentage points or fewer.

[a]Offenses included in our immigration category include both criminal immigration offenses and civil immigration violations— administrative grounds of removability.

[b]Numbers may not sum to totals because of rounding. Percentages may not sum to 100 due to rounding.

SCAAP Criminal Alien Population

Based on our random sample of SCAAP criminal aliens incarcerated in state prisons and local jails during fiscal years 2010 through 2015, we estimated that our state and local study population of about 533,000 SCAAP criminal aliens accounted for about 760,100 arrests/transfers by federal arresting agencies. This same population accounted for about 2.7 million arrests/transfers by state or local arresting agencies from 1964 through 2017.[120] We estimated that 92 percent of the arrests/transfers by federal arresting agencies and 83 percent of the arrests/transfers by state and local arresting agencies occurred from 2000 through 2017.

We estimated that the approximately 533,000 SCAAP criminal aliens in our state and local study population were arrested/transferred by federal arresting agencies for a total of about 904,500 offenses, attempted or committed.[121]

A single arrest can be for multiple offenses and being arrested for one or more offenses does not necessarily result in prosecution or a conviction of all, or any, of the offenses for which an individual was arrested. Of the approximately 904,500 offenses, we estimated that about 94 percent were related to immigration offenses, as shown above in Table 11.[122] Each offense category in the table may include an attempt or conspiracy to commit the respective offense.

---

[120] These dates, 1964 through 2017, represent the dates of the oldest and newest records in our FBI dataset. Law enforcement entities send arrest information to the FBI on a voluntary basis; as a result, FBI data on arrest history may not include all arrests. We were unable to determine the difference between a new arrest and a transfer from one arresting agency to another; as such, we are reporting on arrests and transfers. Because our sample populations are separate for our federal criminal alien population and SCAAP criminal alien population, results cannot be compared to the results we presented in our 2011 report, see GAO-11-187. See appendix I for more details on the margin of error for these estimates. As stated previously, SCAAP criminal aliens represent a portion of the total population of criminal aliens incarcerated in state prisons and local jails.

[121] For our analysis, (a) multiple counts of the same offense were counted once, (b) duplicate offenses on the same day may be counted more than once if the person was transferred to another agency on the same day and that agency submitted the same offense to the FBI, and (c) duplicate offenses on the same day submitted by the same agency may be counted more than once if there was not enough information to determine that two offenses were the same.

[122] See appendix I for a complete description of each of the offense categories.

Table 11. Estimated Number and Percent of Attempted or Committed Offenses for Which State Criminal Alien Assistance Program (SCAAP) Criminal Aliens Incarcerated in State Prisons and Local Jails from Fiscal Years 2010 through 2015 who had an FBI Number Were Arrested/Transferred by Federal Arresting Agencies from 1972 through 2017

Arrest offense	Estimated number	Estimated percent
Immigration[a]	846,500	93.6
Drugs	27,300	3.0
Fraud, forgery, and counterfeiting	9,800	1.1
Obstruction of justice	6,600	0.7
Homicide	3,300	0.4
Miscellaneous	3,300	0.4
Assault	2,200	0.2
Weapons violations	2,200	0.2
Kidnapping	1,100	0.1
Sex offenses	1,100	0.1
Traffic violations	1,100	0.1
Arson	--	--
Burglary	--	--
Disorderly Conduct	--	--
Larceny/theft	--	--
Motor vehicle theft	--	--
Property damage	--	--
Robbery	--	--
Terrorism	--	--
Total[b]	904,500	100

Legend:-- = no estimated offenses.

Source: GAO analysis of Federal Bureau of Investigation (FBI) data. | GAO-18-433.

Note: Offenses may include an attempt or conspiracy to commit the respective offense. While our analyses allowed us to estimate and provide valuable insights about the arrest history of our study population, which consisted of about 533,000 SCAAP criminal aliens incarcerated in state prisons and local jails from fiscal years 2010 through 2015 with an FBI number, our analyses are not intended to infer conclusions about the arrest history of criminal aliens not in this study population. Law enforcement entities send arrest information to the FBI on a voluntary basis; as a result, FBI data on arrest history may not include all arrests. The numbers in the table above represent the number of offenses we estimated that SCAAP criminal aliens were arrested or transferred for; they do not represent the number of times that SCAAP criminal aliens were arrested or transferred for each offense. All estimates in this table have a margin of error of +/- 3 percentage points or fewer.

[a]Offenses included in our immigration category include both criminal immigration offenses and civil immigration violations— administrative grounds of removability.

[b]Numbers may not sum to totals because of rounding. Percentages may not sum to 100 due to rounding.

Table 12. Estimated Number and Percent of Attempted or Committed Offenses for Which State Criminal Alien Assistance Program (SCAAP) Criminal Aliens Incarcerated in State Prisons and Local Jails from Fiscal Years 2010 through 2015 who had an FBI Number Were Arrested/Transferred by State and Local Arresting Agencies from 1972 through 2017

Arrest offense	Estimated number	Estimated percent
Traffic violations	1,224,900	26.9
Drugs	733,800	16.1
Obstruction of justice	658,400	14.4
Assault	394,800	8.7
Larceny/theft	276,700	6.1
Miscellaneous	253,700	5.6
Fraud, forgery, and counterfeiting	190,300	4.2
Burglary	175,000	3.8
Weapons violations	122,500	2.7
Sex offenses	119,200	2.6
Disorderly Conduct	91,000	2.0
Motor vehicle theft	90,800	2.0
Stolen property	75,500	1.7
Robbery	54,700	1.2
Property damage	50,300	1.1
Homicide	24,100	0.5
Kidnapping	17,500	0.4
Immigration[a]	5,500	0.1
Arson	3,300	0.1
Terrorism	1,100	< 0.1
Total[b]	4,562,700	100

Source: GAO analysis of Federal Bureau of Investigation (FBI) data. | GAO-18-433.

Note: Offenses may include an attempt or conspiracy to commit the respective offense While our analyses allowed us to estimate and provide valuable insights about the arrest history of our study population, which consisted of about 533,000 SCAAP criminal aliens incarcerated in state prisons and local jails from fiscal years 2010 through 2015 with an FBI number, our analyses are not intended to infer conclusions about the arrest history of criminal aliens not in this study population. Law enforcement entities send arrest information to the FBI on a voluntary basis; as a result, FBI data on arrest history may not include all arrests. The numbers in the table above represent the number of offenses we estimated that SCAAP criminal aliens were arrested or transferred for; they do not represent the number of times that SCAAP criminal aliens were arrested or transferred for each offense. All estimates in this table have a margin of error of +/- 4 percentage points or fewer.

[a]Offenses included in our immigration category include both criminal immigration offenses and civil immigration violations— administrative grounds of removability.

[b]Numbers may not sum to totals because of rounding. Percentages may not sum to 100 due to rounding.

We estimated that the approximately 533,000 SCAAP criminal aliens in our state and local study population were arrested/transferred by state and local arresting agencies for a total of about 4.6 million offenses. Of these offenses, we estimated that 43 percent were related to traffic violations or to drug offenses, as shown above in Table 12. Each offense category in the table may include an attempt or conspiracy to commit the respective offense.

APPENDIX IV: FEDERAL CONVICTIONS, FISCAL YEARS 2011 THROUGH 2016

This appendix provides additional details on federal convictions of U.S. citizens and criminal aliens. In fiscal year 2016, immigration (66 percent) or drug offenses (25 percent) accounted for 91 percent of the approximately 28,000 primary offenses for which criminal aliens were convicted and sentenced, based upon U.S. Sentencing Commission data, as shown in Figure 17.[123]

---

[123] These were the most recent data available for our review. The U.S. Sentencing Commission is an independent agency in the judicial branch of government. The Commission's data are limited to felony and Class A misdemeanor cases for offenders who are convicted and sentenced in the federal court system. These data do not include state cases, federal petty offenses, federal cases which result in all charges being dismissed or acquitted, federal death penalty cases, federal juvenile cases, or federal witness protection cases. They also do not include convicted offenders for whom no sentences were yet issued, offenders sentenced but for whom no sentencing documents were submitted to the Commission, and offenders sentenced prior to the enactment of the Sentencing Reform Act of 1984. See Pub. L. No. 98-473, ch. II, 98 Stat. 1837, 1987-2001 (1984). According to U.S. Sentencing Commission officials, data do include a small number of individuals that may have been fined or given probation instead of a federal prison sentence. Convictions that did not have information on offender's citizenship status were excluded. Information on the citizenship status of offenders is obtained from their presentence investigation report. The U.S. Sentencing Commission data on non-U.S. citizens includes data on "resident aliens," "illegal aliens," "extradited aliens" and "non-U.S. citizens, alien status unknown." For the purposes of this report, we refer to these individuals as criminal aliens. Data on primary offenses for which individuals may be convicted include attempts, conspiracies, and intents to commit crimes. For example, the U.S. Sentencing Commission's offense category of assault includes attempt to commit murder and assault with intent to murder; while the murder offense category includes a conspiracy to murder when there is a resulting death. See appendix I for a complete description of each of the offense categories.

These data are consistent with trends from fiscal years 2011 through 2016. These criminal aliens may also be in our study population of criminal aliens incarcerated in federal prisons if they were convicted in federal court and sentenced during fiscal years 2011 through 2016, and were also incarcerated in a federal prison during this same time period.

Of selected immigration offenses for which convicted criminal aliens were sentenced in fiscal year 2016, 91 percent were for illegal reentry, as shown in Figure 18.[124]

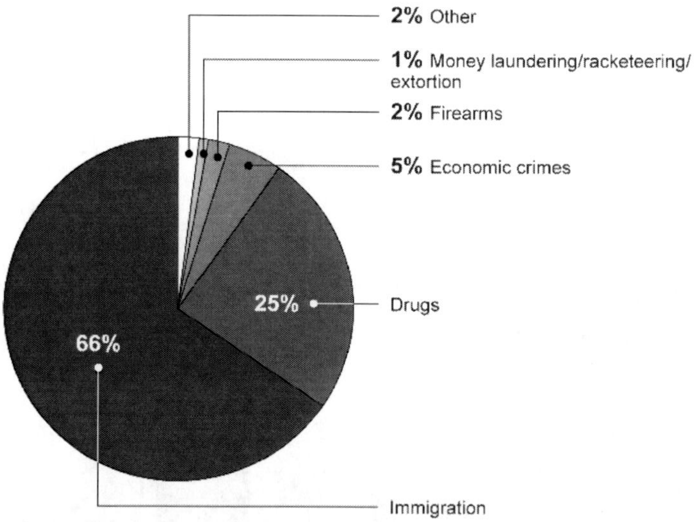

Source: GAO analysis of U.S. Sentencing Commission data. | GAO-18-433.

Note: "Other" includes administration of justice offenses, sex offenses, assault, robbery, homicide, auto theft, kidnapping/hostage taking, arson, burglary, and other miscellaneous offenses.

Figure 17. Primary Offense Category for Which Criminal Aliens Were Convicted in Federal Court and Sentenced in Fiscal Year 2016.

---

[124] These selected immigration offenses accounted for 95 percent of all immigration offenses for which criminal aliens were convicted of in fiscal year 2016. According to the U.S. Sentencing Commission, subsequent entries, reentry after removal, and remaining in the United States after being ordered removed are felonies covered by U.S. Sentencing Guidelines and as such would be included in these data. Further, conviction for a first offense of illegal entry is a misdemeanor and would not be covered by U.S. Sentencing Guidelines in these data. See 8 U.S.C. §§ 1325(a), 1326, and 1253.

For the primary offenses for which all individuals—U.S. citizens and criminal aliens—were convicted and sentenced in fiscal years 2011 through 2016, criminal aliens made up between 0 and 94 percent of all convictions for each primary offense, based upon U.S. Sentencing Commission data, as shown in Table 13.

Table 14 details the total number of primary offenses for which U.S. citizens and criminal aliens were convicted and sentenced in each fiscal year from 2011 through 2016.

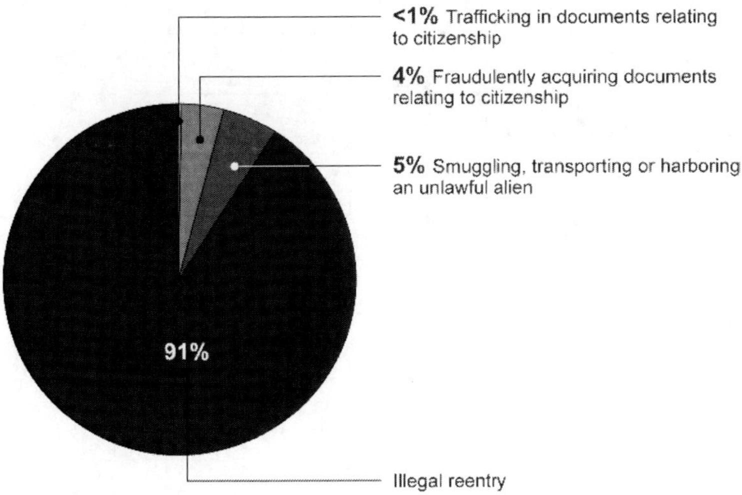

Source: GAO analysis of U.S. Sentencing Commission data. | GAO-18-433.

Note: These selected immigration offenses accounted for 95 percent of all immigration offenses for which criminal aliens were convicted in federal court and sentenced in fiscal year 2016.

Figure 18. Selected Immigration Primary Offenses for Which Convicted Criminal Alien Federal Offenders Were Sentenced in Fiscal Year 2016.

Primary offenses related to drugs, immigration, economic crimes, and firearms accounted for 87 percent of all federal convictions of individuals that were sentenced in fiscal year 2011 through fiscal year 2016. Figure 19 shows the trends for each of these primary offenses.

Table 13. Number of Each Primary Offense for Federal Convictions, by Citizenship, Fiscal Years 2011 through 2016

Primary offense	Total, all convictions	Total, U.S. citizen	Percentage of all convictions that were for U.S. citizens	Total, criminal aliens	Percentage of all convictions that were for criminal aliens
Drugs	143,899	98,582	68.5	45,317	31.5
Immigration	143,448	8,739	6.1	134,709	93.9
Economic crimes	62,063	51,026	82.2	11,037	17.8
Firearms	46,198	42,712	92.5	3,486	7.5
Sex offenses	14,488	13,897	95.9	591	4.1
Money laundering/ racketeering/extortion	10,011	7,819	78.1	2,192	21.9
Administration of justice offenses	7,183	6,217	86.6	966	13.4
Robbery	4,867	4,708	96.7	159	3.3
Assault	4,235	3,827	90.4	408	9.6
Homicide	811	739	91.1	72	8.9
Auto theft	423	369	87.2	54	12.8
Arson	330	315	95.5	15	4.5
Kidnapping/hostage taking	290	167	57.6	123	42.4
Burglary/breaking and entering	227	227	100.0	0	0.0
Other	16,977	14,776	87.0	2,201	13.0
Total	455,450	254,120	n/a	201,330	n/a

Legend: n/a = not applicable. Source: GAO Analysis of U.S. Sentencing Commission data. | GAO-18-433.

Note: U.S. Sentencing Commission data are limited to felony and Class A misdemeanor cases for offenders who are convicted and sentenced in the federal court system. These data do not include state cases, federal petty offenses, federal cases which result in all charges being dismissed or acquitted, federal death penalty cases, federal juvenile cases, or federal witness protection cases. They also do not include convicted offenders for whom no sentences were yet issued, offenders sentenced but for whom no sentencing documents were submitted to the Commission, and offenders sentenced prior to the enactment of the Sentencing Reform Act of 1984. See Pub. L. No. 98-473, ch. II, 98 Stat. 1837, 1987-2001 (1984). According to U.S. Sentencing Commission officials, data do include a small number of individuals that may have been fined or given probation instead of a federal prison sentence. Convictions that did not have information on offender's citizenship status were excluded. Information on the citizenship status of offenders is obtained from their presentence investigation report. The U.S. Sentencing Commission data on non-U.S. citizens includes data on "resident aliens," "illegal aliens," "extradited aliens" and "non-U.S. citizens, alien status unknown." For the purposes of this chapter, we refer to these individuals as criminal aliens. Data on primary offenses for which individuals may be convicted include attempts, conspiracies, and intents to commit crimes. For example, the U.S. Sentencing Commission's offense category of assault includes attempt to commit murder and assault with intent to murder; while the murder offense category includes a conspiracy to murder when there is a resulting death. The "other" category includes bribery, gambling/lottery, civil rights, prison offenses, environmental/wildlife, national defense, food and drug, and other miscellaneous offenses.

Table 14. Number of Each Primary Offense for Federal Convictions, by Citizenship, Fiscal Years 2011 through 2016

	2011	2012	2013	2014	2015	2016	5-year change
U.S. citizens							
Primary offense							
Drugs	17,616	17,589	16,840	16,568	15,562	14,407	-18
Immigration	1,579	1,283	1,300	1,381	1,517	1,679	6
Economic crimes	8,873	8,974	8,612	9,029	8,076	7,462	-16
Firearms	7,041	7,371	7,420	7,380	6,627	6,873	-2
Sex offenses	2,179	2,358	2,238	2,352	2,325	2,445	12
Money laundering/ racketeering/ extortion	1,162	1,318	1,306	1,374	1,375	1,284	10
Administration of justice offenses	951	1,105	1,187	1,058	1,005	911	-4
Robbery	914	824	809	759	723	679	-26
Assault	568	585	597	684	672	721	27
Homicide	131	108	142	113	126	119	-9
Auto theft	71	50	83	72	51	42	-41
Arson	53	46	66	50	55	45	-15
Kidnapping/ hostage taking	25	25	28	30	22	37	48
Burglary/breakin g and entering	52	42	37	37	36	23	-56
Other offenses	2,338	2,183	2,311	2,592	2,887	2,465	5
Total	43,553	43,861	42,976	43,479	41,059	39,192	-10
Criminal aliens							
Primary offense							
Drugs	7,823	8,518	7,782	7,347	6,923	6,924	-11
Immigration	27,912	24,879	23,496	20,823	19,240	18,359	-34
Economic crimes	2,498	2,187	1,900	1,550	1,575	1,327	-47
Firearms	783	706	618	532	421	426	-46
Sex offenses	70	83	105	118	105	110	57
Money laundering/ racketeering/ extortion	396	400	384	370	288	354	-11

Table 14. (Continued)

	2011	2012	2013	2014	2015	2016	5-year change
Administration of justice offenses	185	177	164	171	157	112	-39
Robbery	41	25	32	23	15	23	-44
Assault	58	71	79	69	62	69	19
Homicide	8	6	8	11	17	22	175
Auto theft	13	10	3	14	11	3	-77
Arson	1	2	3	4	4	1	0
Kidnapping/ hostage taking	44	24	12	20	21	2	-95
Burglary/breakin g and entering	0	0	0	0	0	0	0
Other	332	424	460	380	327	278	-16
Total	40,164	37,512	35,046	31,432	29,166	28,010	-30

Source: GAO Analysis of U.S. Sentencing Commission data. | GAO-18-433.

Note: U.S. Sentencing Commission data are limited to felony and Class A misdemeanor cases for offenders who are convicted and sentenced in the federal court system. These data do not include state cases, federal petty offenses, federal cases which result in all charges being dismissed or acquitted, federal death penalty cases, federal juvenile cases, or federal witness protection cases. They also do not include convicted offenders for whom no sentences were yet issued, offenders sentenced but for whom no sentencing documents were submitted to the Commission, and offenders sentenced prior to the enactment of the Sentencing Reform Act of 1984. See Pub. L. No. 98-473, ch. II, 98 Stat. 1837, 1987-2001 (1984). According to U.S. Sentencing Commission officials, data do include a small number of individuals that may have been fined or given probation instead of a federal prison sentence. Convictions that did not have information on offender's citizenship status were excluded. Information on the citizenship status of offenders is obtained from their presentence investigation report. The U.S. Sentencing Commission data on non-U.S. citizens includes data on "resident aliens," "illegal aliens," "extradited aliens" and "non-U.S. citizens, alien status unknown." For the purposes of this chapter, we refer to these individuals as criminal aliens. Data on primary offenses for which individuals may be convicted include attempts, conspiracies, and intents to commit crimes. For example, the U.S. Sentencing Commission's offense category of assault includes attempt to commit murder and assault with intent to murder; while the murder offense category includes a conspiracy to murder when there is a resulting death. The "other" category includes bribery, gambling/lottery, civil rights, prison offenses, environmental/wildlife, national defense, food and drug, and other miscellaneous offenses.

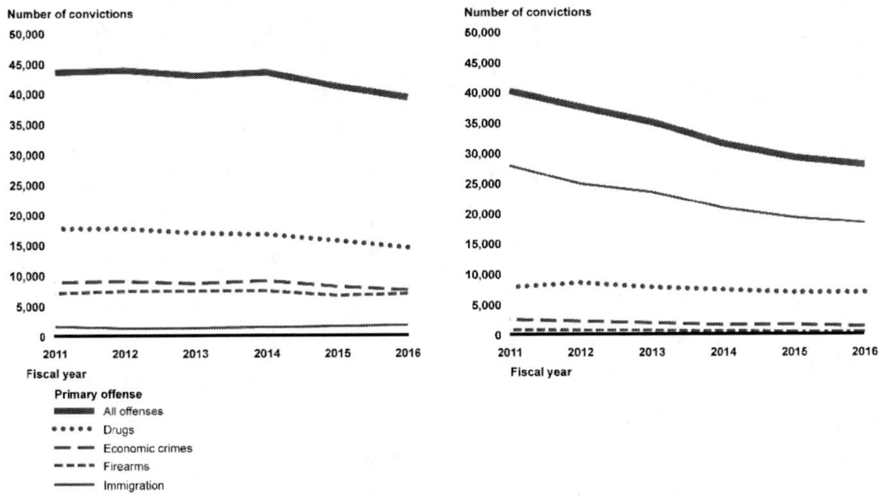

Source: GAO Analysis of U.S. Sentencing Commission data. | GAO-18-433.

Note: U.S. Sentencing Commission data are limited to felony and Class A misdemeanor cases for offenders who are convicted and sentenced in the federal court system. These data do not include state cases, federal petty offenses, federal cases which result in all charges being dismissed or acquitted, federal death penalty cases, federal juvenile cases, or federal witness protection cases. They also do not include convicted offenders for whom no sentences were yet issued, offenders sentenced but for whom no sentencing documents were submitted to the Commission, and offenders sentenced prior to the enactment of the Sentencing Reform Act of 1984. See Pub. L. No. 98-473, ch. II, 98 Stat. 1837, 1987-2001 (1984). According to U.S. Sentencing Commission officials, data do include a small number of individuals that may have been fined or given probation instead of a federal prison sentence. Convictions that did not have information on offender's citizenship status were excluded. Information on the citizenship status of offenders is obtained from their presentence investigation report. The U.S. Sentencing Commission data on non-U.S. citizens includes data on "resident aliens," "illegal aliens," "extradited aliens" and "non-U.S. citizens, alien status unknown." For the purposes of this chapter, we refer to these individuals as criminal aliens.

Figure 19. Primary Offenses for Which Individuals Were Convicted, for Offense Categories with the Greatest Number of Federal Convictions, by Citizenship, from Fiscal Years 2011 through 2016.

APPENDIX V: INDIVIDUALS CONVICTED AS A RESULT OF INTERNATIONAL TERRORISM-RELATED INVESTIGATIONS

The Department of Justice's (DOJ) National Security Division (NSD) maintains a list of individuals with public and unsealed federal convictions resulting from international terrorism-related investigations conducted since September 11, 2001.[125] According to DOJ, the list includes both individuals convicted of crimes that DOJ considers to be directly related to international terrorism and individuals convicted of a variety of other crimes where the investigation, at the time of charging, appeared to involve an identified link to international terrorism.[126] According to the U.S. Attorney's Manual, U.S. Attorneys are to coordinate with the NSD on all investigations involving an identified link to international terrorism. According to DOJ officials, when such investigations result in a public and unsealed criminal conviction, the defendant's name and information about the charged crime and conviction, among other information, is added to the NSD list. According to DOJ, the NSD list does not include convictions

---

[125] DOJ, *National Security Division Chart of Public/Unsealed Terrorism and Terrorism-Related Convictions, 9/11/01-12/31/16.* (Washington, D.C.: February 10, 2017). For the purposes of this report, we refer to this chart as the NSD list. According to the NSD, these investigations include investigations of terrorist acts planned or committed outside the territorial jurisdiction of the United States over which Federal criminal jurisdiction exists and investigations of terrorist acts planned or committed within the United States involving international terrorists and terrorist groups. Individuals whose convictions arose from the nationwide investigation conducted after the September 11, 2001 investigation were included on the list at that time, regardless of whether investigators developed or identified evidence that they had any connection to international terrorism. Since then, the NSD has added individuals to this list only if, at the time of charging, they appeared to have a connection to international terrorism, even if they were not charged with a terrorism offense. The decision to add defendants to the list is made on a case-by-case basis by career prosecutors in the NSD's Counterterrorism Section, whose primary responsibility is investigating and prosecuting international and domestic terrorism cases to prevent and disrupt acts of terrorism anywhere in the world that impact on significant United States interests and persons. According to the NSD, its list is distinct from statistics maintained by the Bureau of Prisons to track inmates with terrorist connections.

[126] In January 2018, DOJ and the Department of Homeland Security issued a report that included information about some individuals on the NSD list. Department of Homeland Security and Department of Justice, *Executive Order 13780: Protecting the Nation From Foreign Terrorist Entry Into the United States Initial Section 11 Report* (January 2018).

related solely to domestic terrorism. We did not verify the connections to terrorism for individuals on the NSD list.

There are 250 individuals on the NSD list with convictions from March 19, 2010 through December 2016.[127] Of these, 196 individuals (or 78 percent) were convicted under statutes NSD considers directly related to international terrorism—regardless of the offense for which they were charged.[128] For example, some individuals on the NSD list were convicted of providing material support to designated terrorist organizations, which are designated by the Secretary of State and include, among others, Boko Haram, Hamas, al Qaeda, and the Revolutionary Armed Forces of Colombia.[129] We used data from the Bureau of Prisons, U.S. Sentencing Commission, U.S. Citizenship and Immigration Services, and U.S. Immigration and Customs Enforcement to determine the citizenship status of individuals convicted under statutes NSD considers directly related to international terrorism from March 2010 through December 2016 at the time of their conviction, as shown in Table 15.[130]

---

[127] In 2011, we reported information on the 399 individuals on the NSD list with convictions between September 2001 and March 18, 2010. See GAO, *Criminal Alien Statistics: Information on Incarcerations, Arrests, and Costs,* GAO-11-187 (Washington, D.C.: Mar. 24, 2011).

[128] As defined by DOJ's NSD, criminal cases arising from international terrorism investigations are divided into two categories. The first category includes cases with charged violations of federal statutes that are directly related to international terrorism, regardless of the offense of conviction. These statutes prohibit terrorist acts abroad against U.S. nationals and providing material support to terrorists, among others. The second category includes cases charged with a variety of other statutes where the investigation involved an identified link to international terrorism. These include offenses such as those involving fraud, immigration, firearms, drugs, and perjury, among others.

[129] The Secretary of State, in consultation with the Secretary of the Treasury and the Attorney General, has the authority to designate a foreign organization as a foreign terrorist organization. Designation allows the United States to impose legal consequences on the foreign terrorist organization or on individuals who support the foreign terrorist organization. See GAO, *Combating Terrorism: Foreign Terrorist Organization Designation Process and U.S. Agency Enforcement Actions,* GAO-15-629 (Washington, D.C.: June 25, 2015).

[130] We relied on the U.S. Sentencing Commission data to identify those aliens who were extradited to the United States for prosecution. We relied on U.S. Citizenship and Immigration Services data to identify naturalized citizens. We determined the citizenship status of the remaining individuals on the list using the totality of available data from these agencies.

Table 15. Citizenship Status of Individuals Convicted under Statutes Directly Related to International Terrorism, March 2010 through December 2016

Citizenship status	Number of individuals convicted under statute directly related to terrorism	Percent
Alien	95	48
Extradited alien[a]	*31*	--
Other alien	*64*	--
U.S. citizen[b]	97	49
Naturalized U.S. citizen[c]	*24*	--
U.S. citizen, not naturalized	*73*	--
Unknown	4	2
Total[d]	196	100

Legend: -- = percent for sub-categories not provided.

Source: GAO analysis of Bureau of Prisons, U.S. Sentencing Commission, U.S. Citizenship and Immigration Services (USCIS), and U.S. Immigration and Customs Enforcement data. | GAO-18-433.

[a]We relied on U.S. Sentencing Commission data to identify aliens who were brought to the United States for prosecution.

[b]This category includes U.S. citizens who were born in the United States, derived U.S. citizenship, or were naturalized.

[c]We relied on USCIS data to identify naturalized citizens. All individuals identified as naturalized citizens in this table became U.S. citizens before they were charged with a crime related to international terrorism.

[d]Percentages may not add to 100 due to rounding.

Of those 196 individuals with convictions directly related to international terrorism, 95 were aliens, including 31 aliens that were brought to the United States for prosecution. Of the individuals born outside of the United States—including aliens, extradited aliens, U.S. citizens, and individuals whose citizenship status is unknown but whose country of birth is known—convicted of statutes directly related to international terrorism, the highest number of convictions were from individuals born in Somalia (19 convictions), Pakistan (14 convictions), and Colombia (12 convictions).[131] No other country of birth outside of the

[131] Some individuals born in these countries were extradited to the United States for prosecution, including those born in Somalia (3 individuals), Pakistan (4 individuals), and Colombia (10 individuals).

United States had more than five individuals with convictions directly related to international terrorism.

Appendix VI: Primary Offenses for Which Criminal Aliens Incarcerated in Selected State Prison Systems Were Convicted

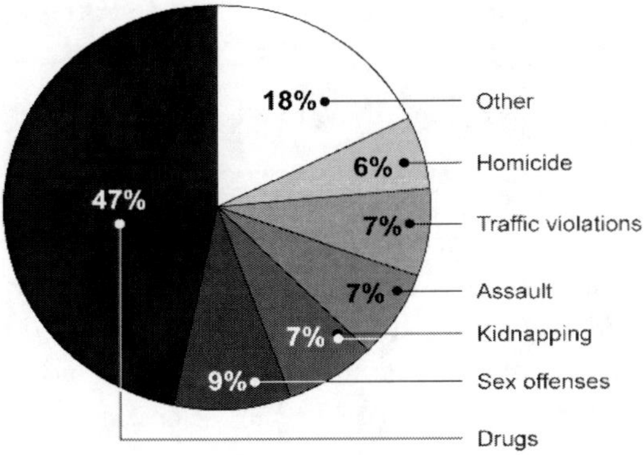

Source: GAO analysis of Arizona Department of Corrections data. | GAO-18-433.

Note: Our analysis is of the primary offense per SCAAP criminal alien incarceration in fiscal year 2015; convictions may have occurred prior to this time period. The figure above includes over 6,300 primary offenses. "Other" includes primary offenses for robbery; burglary; fraud, forgery, and counterfeiting; larceny and theft; motor vehicle theft; obstruction of justice; immigration; stolen property; disorderly conduct; arson; property damage; and miscellaneous offenses. Each of these primary "other" offenses make up between 4.1 percent and .1 percent of all the offenses for which these aliens were convicted. Arizona Department of Corrections determined the primary offense by the sentence which has the controlling release date. Offenses for which individuals are convicted may include an attempt or conspiracy to commit the respective offense. We did not examine the extent to which state policies may have affected the number of individuals convicted of crimes.

Figure 20. Primary Offenses, Attempted or Committed, for Which State Criminal Alien Assistance Program (SCAAP) Criminal Aliens Incarcerated in Arizona's State Prison System in Fiscal Year 2015 Were Convicted.

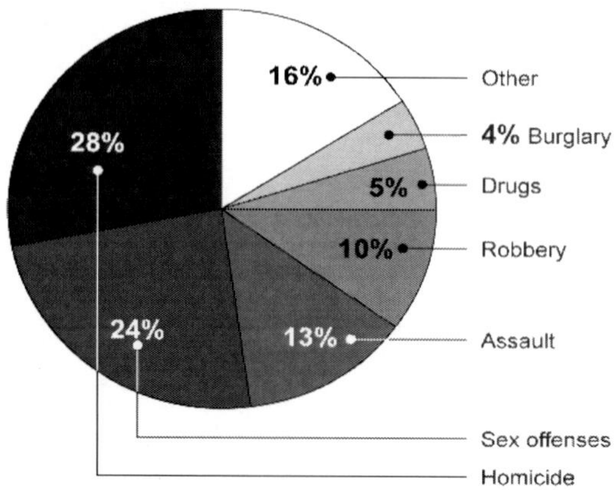

Source: GAO analysis of California Department of Corrections and Rehabilitation data. | GAO-18-433.

Note: Our analysis is of the primary offense per SCAAP criminal alien incarceration in fiscal year 2015; these convictions may have occurred prior to this time period. The figure above includes over 18,600 primary offenses. "Other" includes primary offenses for larceny and theft; traffic violations; fraud, forgery, and counterfeiting; weapons violations; motor vehicle theft; obstruction of justice; stolen property; arson; property damage; and miscellaneous offenses. Each of these primary "other" offenses make up between 2.8 percent and .2 percent of all the offenses for which these aliens were convicted California Department of Corrections and Rehabilitation determines a primary offense based on the longest maximum sentence and/or offense severity. Offenses for which individuals are convicted may include an attempt or conspiracy to commit the respective offense. We did not examine the extent to which state policies may have affected the number of individuals convicted of crimes.

Figure 21. Primary Offenses, Attempted or Committed, for Which State Criminal Alien Assistance Program (SCAAP) Criminal Aliens Incarcerated in California's State Prison System in Fiscal Year 2015 Were Convicted.

Since there are no reliable data on criminal aliens incarcerated in all state prisons and local jails, we analyzed conviction data from the five state prison systems that had the largest number of State Criminal Alien Assistance Program (SCAAP) criminal alien incarcerations in fiscal year 2015.[132] Primary offenses for which SCAAP criminal aliens incarcerated in

[132] These data were the most recent data available. Collectively, these five state prison systems accounted for 64 percent of the SCAAP criminal alien incarcerations in state prisons during

fiscal year 2015 were convicted varied across these selected state prison systems, as shown in Figures 20 through 24.

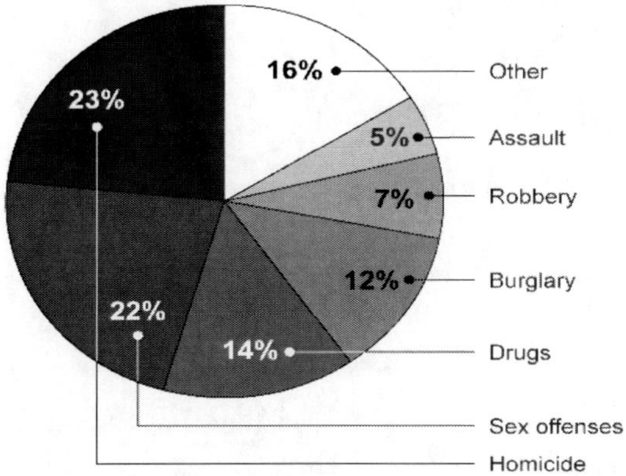

Source: GAO analysis of Florida Department of Corrections data. | GAO-18-433.

Note: Our analysis is of the primary offense per SCAAP criminal alien incarceration in fiscal year 2015; these convictions may have occurred prior to this time period. The figure above includes nearly 6,300 primary offenses. "Other" includes primary offenses for motor vehicle theft; traffic violations; obstruction of justice; kidnapping; weapons violations; larceny and theft; property damage; fraud, forgery, and counterfeiting; stolen property; arson; and miscellaneous offenses. Each of these primary "other" offenses make up between 3.6 percent and .05 percent of all the offenses for which these aliens were convicted. Florida Department of Corrections defines a primary offense as the offense with the highest felony degree. If there multiple offenses with the same felony degree, the primary offense is the one with the longest sentence. If there are multiple offenses with the same felony degree and sentence length, the first offense listed on the commitment papers is the primary offense. Offenses for which individuals are convicted may include an attempt or conspiracy to commit the respective offense. We did not examine the extent to which state policies may have affected the number of individuals convicted of crimes.

Figure 22. Primary Offenses, Attempted or Committed, for Which State Criminal Alien Assistance Program (SCAAP) Criminal Aliens Incarcerated in Florida's State Prison System in Fiscal Year 2015 Were Convicted.

---

fiscal year 2015. This analysis included state prison systems that participated in SCAAP and did not include U.S. territories. State prison systems in Arkansas, West Virginia, Vermont, and the District of Columbia did not receive reimbursement for SCAAP criminal aliens incarcerated in fiscal year 2015.

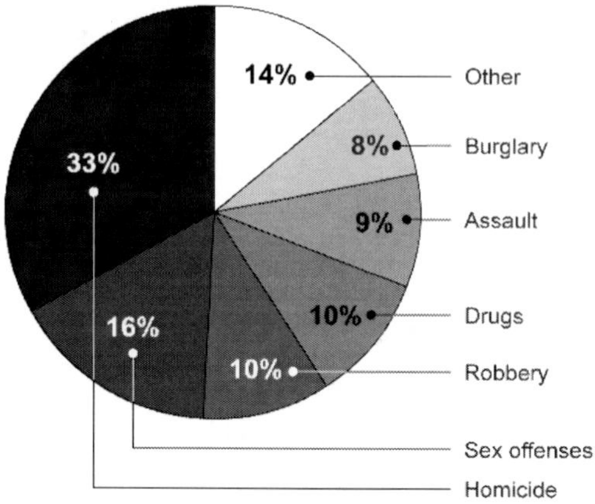

Source: GAO analysis of New York State Department of Corrections and Community Supervision data. | GAO-18-433.

Note: Our analysis is of the primary offense per SCAAP criminal alien incarceration in fiscal year 2015; these convictions may have occurred prior to this time period. The figure above includes nearly 3,400 primary offenses. "Other" includes primary offenses for weapons violations; kidnapping; larceny and theft; traffic violations; fraud, forgery, and counterfeiting; arson; stolen property; obstruction of justice; and miscellaneous offenses. Each of these primary "other" offenses make up between 4.2 percent and .2 percent of all the offenses for which these aliens were convicted. New York State Department of Corrections and Community Supervision defines a primary offense in almost all instances as the offense with the longest sentence. In some cases, a violent felony may become the primary offense even if the sentence for that crime is not the longest. Offenses for which individuals are convicted may include an attempt or conspiracy to commit the respective offense. We did not examine the extent to which state policies may have affected the number of individuals convicted of crimes.

Figure 23. Primary Offenses, Attempted or Committed, for Which State Criminal Alien Assistance Program (SCAAP) Criminal Aliens Incarcerated in New York's State Prison System in Fiscal Year 2015 Were Convicted.

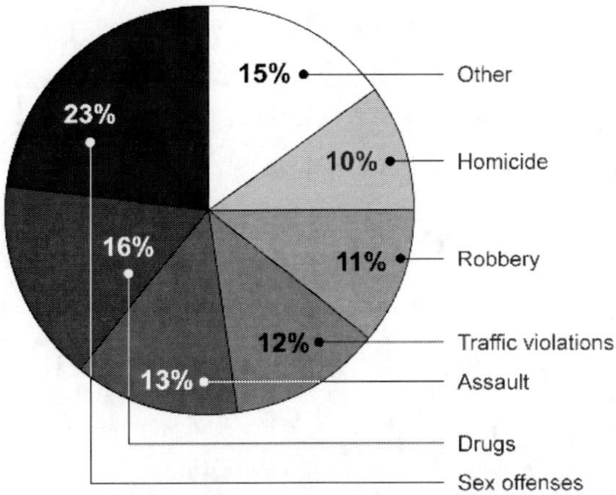

Source: GAO analysis of Texas Department of Criminal Justice data. | GAO-18-433.

Note: Our analysis is of the primary offense per SCAAP criminal alien incarceration in fiscal year 2015; these convictions may have occurred prior to this time period. The figure above includes nearly 9,600 primary offenses. "Other" includes primary offenses for burglary; obstruction of justice; kidnapping; larceny and theft; fraud, forgery, and counterfeiting; weapons violations; motor vehicle theft; arson; property damage; stolen property; disorderly conduct; and miscellaneous offenses. Each of these primary "other" offenses make up between 5 percent and .01 percent of all the offenses for which these aliens were convicted. Texas Department of Criminal Justice defines the primary offense as the offense that keeps the offender incarcerated the longest if there is more than one current offense. Offenses for which individuals are convicted may include an attempt or conspiracy to commit the respective offense. We did not examine the extent to which state policies may have affected the number of individuals convicted of crimes.

Figure 24. Primary Offenses, Attempted or Committed, for Which State Criminal Alien Assistance Program (SCAAP) Criminal Aliens Incarcerated in Texas's State Prison System in Fiscal Year 2015 Were Convicted.

Appendix VII: Estimated Costs to Incarcerate Criminal Aliens in Fiscal Year 2016 Dollars

This appendix provides cost data for the following: (1) estimated federal costs to incarcerate criminal aliens; (2) SCAAP reimbursements to states and localities, and (3) estimated selected operating costs to incarcerate state SCAAP criminal aliens in all 50 state prison systems from fiscal years 2010 through 2015. These cost data are presented in fiscal year 2016 dollars, as shown in Figures 25 through 27.

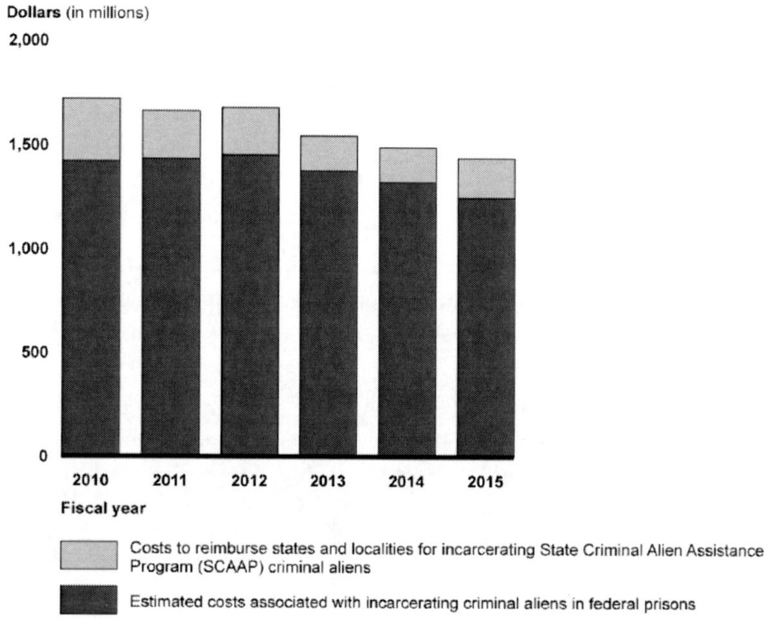

Source: GAO analysis of Bureau of Prisons and Bureau of Justice Assistance data. | GAO-18-433.

Note: Costs to reimburse states and localities are based on actual SCAAP reimbursements each year. Costs associated with incarcerating criminal aliens in federal prisons were estimated based on snapshot data representing an average of the 12 monthly population snapshots for each type of Bureau of Prisons institution, and Bureau of Prisons per capita costs.

Figure 25. Estimated Federal Costs to Incarcerate Criminal Aliens from Fiscal Years 2010 through 2015 in Fiscal Year 2016 Dollars.

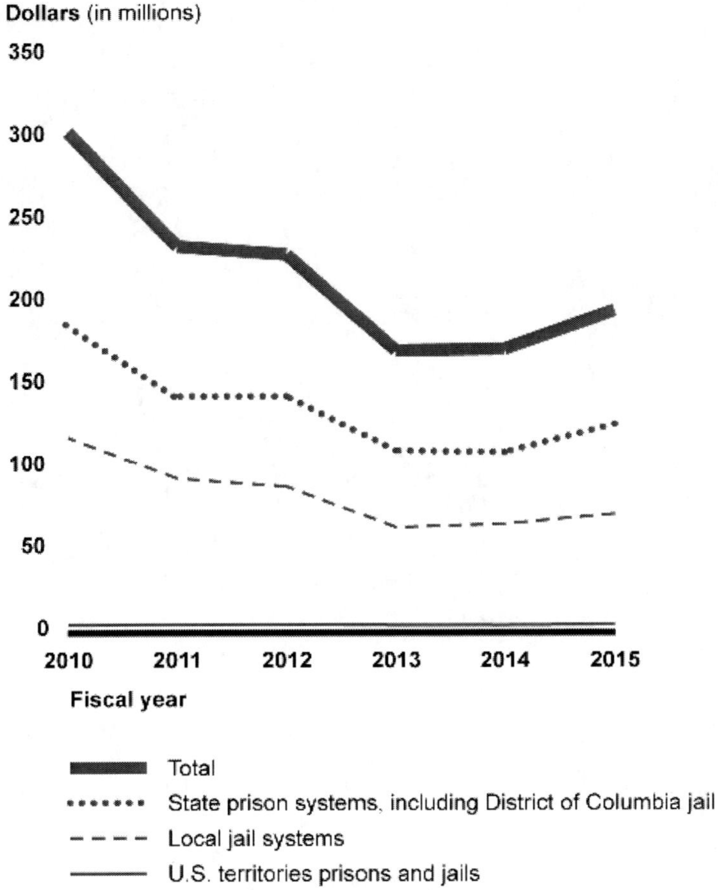

Source: GAO analysis of Bureau of Justice Assistance data. | GAO-18-433.

Note: Forty-seven state prison systems and four U.S. territories received SCAAP reimbursement in fiscal year 2015 compared to fifty state prison systems, including the District of Columbia, and four U.S. territories in fiscal year 2010. 760 local jail systems received reimbursement in fiscal year 2015 compared to 875 local jail systems in fiscal year 2010.

Figure 26. State Criminal Alien Assistance Program (SCAAP) Reimbursements to States and Localities from Fiscal Years 2010 through 2015 in Fiscal Year 2016 Dollars.

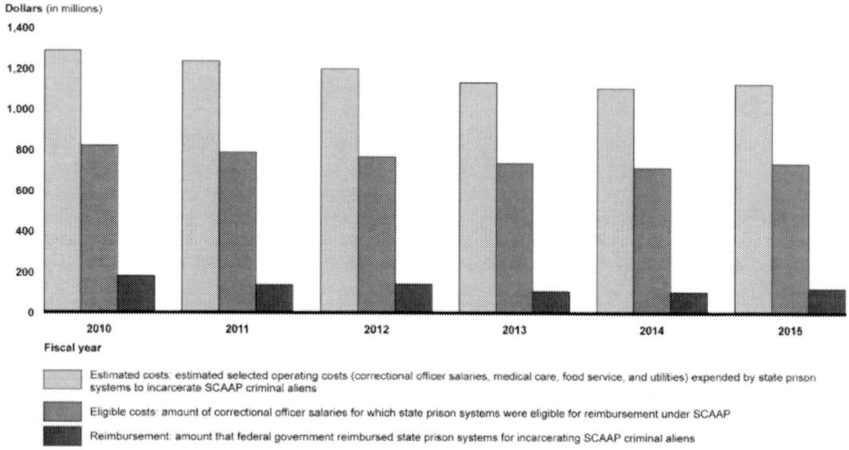

Source: GAO analysis of Bureau of Justice Statistics and Bureau of Justice Assistance data. | GAO-18-433.

Note: SCAAP reimbursement figures may not equal appropriation due to rounding. Our analysis includes those state prison systems—including the District of Columbia jail— that received SCAAP reimbursement in the associated fiscal year. As a result, not all 50 state prison systems may be included in these estimates for each fiscal year.

Figure 27. Estimated Selected Operating Costs to Incarcerate State Criminal Alien Assistance Program (SCAAP) Criminal Aliens in All 50 State Prison Systems from Fiscal Years 2010 through 2015 in Fiscal Year 2016 Dollars.

Appendix VIII: Estimated Costs and Federal Reimbursements to Incarcerate Criminal Aliens in Selected States and Localities

Based on average daily cost data, we estimated the total costs expended by five selected state prion systems to incarcerate State Criminal Alien Assistance Program (SCAAP) criminal aliens in fiscal year 2015. Specifically, as shown in Figure 28, the federal government reimbursed the state prison systems with the largest number of SCAAP criminal alien incarcerations—Arizona, California, Florida, New York, and Texas—only a portion of their estimated costs to incarcerate SCAAP criminal aliens in state prisons.[133]

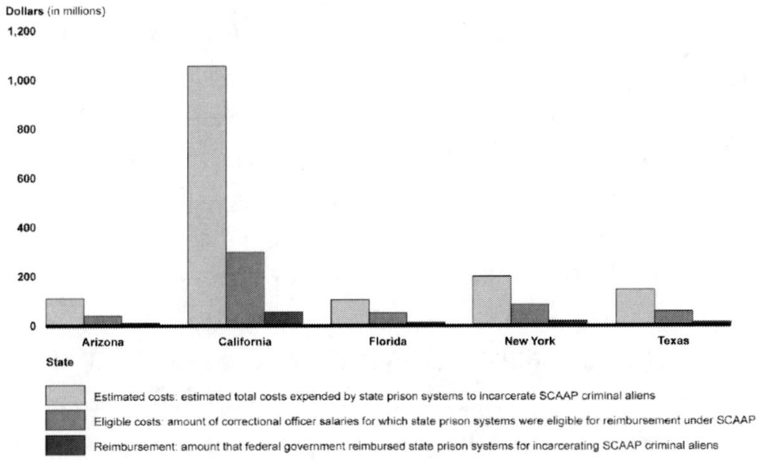

Source: GAO analysis of Bureau of Justice Assistance data and Arizona Department of Corrections, California Department of Corrections and Rehabilitation, Florida Department of Corrections, New York State Department of Corrections and Community Supervision, and Texas Department of Criminal Justice data. | GAO-18-433.

Note: SCAAP reimbursement figures may not equal the actual SCAAP awarded amount due to rounding.

Figure 28. Selected State Prison Systems' Estimated Costs and Federal Reimbursements to Incarcerate State Criminal Alien Assistance Program (SCAAP) Criminal Aliens in Fiscal Year 2015.

Specifically, for these selected state prison systems, the federal government's reimbursement, which is only for a portion of correctional officer salaries, accounted for between 5 and 8 percent of the estimated costs associated with incarcerating SCAAP criminal aliens in fiscal year

---

[133] We selected the five state prison systems with the highest number of SCAAP criminal alien incarcerations. Collectively, these state prison systems accounted for 64 percent of the SCAAP criminal alien incarcerations in state prisons during fiscal year 2015. These same state prison systems were selected in our April 2005 and March 2011 report. See GAO, *Criminal Alien Statistics: Information on Incarcerations, Arrests, and Costs,* GAO-11-187 (Washington, D.C.: Mar. 24, 2011); *Information on Criminal Aliens Incarcerated in Federal and State Prisons and Local Jails*, GAO-05-337R (Washington, D.C.: Apr. 7, 2005).

2015.[134] We estimated the total costs that each of the five state prison systems expended to incarcerate SCAAP criminal aliens—based on their average daily costs—totaled about $1.6 billion in 2015, with California accounting for 66 percent of these costs. Of the estimated $1.6 billion, these five state prison systems were eligible to be reimbursed for about $509 million in correctional officer salaries under SCAAP. Based on available appropriations for SCAAP for fiscal year 2015, which reimbursed states and localities for 17 percent of eligible costs, the federal government reimbursed these state prison systems about $86.5 million of the approximately $509 million they were eligible to receive.

Based on average daily cost data, we estimated the total costs expended by five selected local jail systems to incarcerate SCAAP criminal aliens in fiscal year 2015. Specifically, as shown in Figure 29, the federal government also reimbursed the local jail systems with the largest number of SCAAP criminal alien incarcerations—Maricopa County, Arizona; Orange County, California; Los Angeles County, California; Essex County, New Jersey; and Harris County, Texas—only a portion of their estimated expenditures to incarcerate SCAAP criminal aliens in local jails.[135]

Specifically, for these selected local jail systems, the federal government's reimbursement, which is only for a portion of correctional officer salaries, accounted for between 4 and 7 percent of the estimated costs associated with incarcerating SCAAP criminal aliens in fiscal year

---

[134] Our estimates of the total costs for incarcerating SCAAP criminal aliens in state prisons are based on each jurisdiction's average daily cost data and number of inmate days attributed to SCAAP criminal aliens. These average daily costs might not represent actual expenditures if expenditures on undocumented prisoners differ from expenditures in the average prison population.

[135] We selected the six local jail systems with the highest number of SCAAP criminal alien incarcerations. Collectively, these local jail systems accounted for 19 percent of the SCAAP criminal alien incarcerations in local jails during fiscal year 2015. These local jail systems were the same as the ones we selected in our April 2005 and March 2011 report with the exception of Essex County, New Jersey. Additionally, although New York City, New York was in the top six, we could not estimate total costs for this locality. Officials from this locality stated that they no longer apply for SCAAP funds, and they did not provide us an average daily cost per inmate.

2015.[136] We estimated the total costs that each of the five local jail systems expended to incarcerate SCAAP criminal aliens—based on their average daily costs—totaled about $198 million in fiscal year 2015, with Los Angeles, California accounting for 44 percent of these costs. Of the estimated $198 million, these five local jail systems were eligible to be reimbursed for about $66 million in correctional officer salaries under SCAAP. Based on available appropriations for SCAAP for fiscal year 2015 which reimbursed states and localities for 17 percent of eligible costs, the federal government reimbursed these local jail systems about $11.2 million of the approximately $66 million they were eligible to receive.

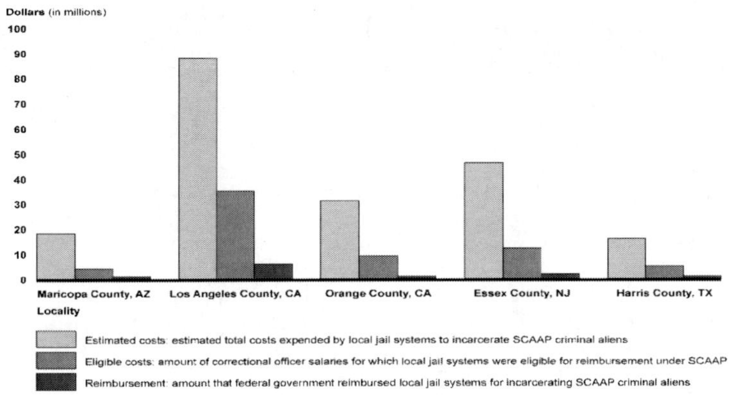

Source: GAO analysis of Bureau of Justice Assistance data and Essex Country, New Jersey Department of Corrections, Los Angeles County, California, Sheriff's Department, Maricopa County, Arizona Sheriff's Office, Orange County, California Sheriff's Department; and Harris County, Texas Sheriff's Office data. | GAO-18-433.

Note: SCAAP reimbursement figures may not equal the actual SCAAP awarded amount due to rounding.

Figure 29. Selected Local Jail Systems' Estimated Costs and Federal Reimbursements for Incarcerating State Criminal Alien Assistance Program (SCAAP) Criminal Aliens in Fiscal Year 2015.

---

[136] Our estimates of the total costs for incarcerating SCAAP criminal aliens in local jails are based on each local jail system's average daily cost data and number of inmate days attributed to SCAAP criminal aliens. These average daily costs might not represent actual expenditures if expenditures on undocumented prisoners differ from expenditures in the average prison population.

APPENDIX IX: REMOVALS FROM THE UNITED STATES OF ALIENS CONVICTED OF A CRIME, FISCAL YEARS 2011 THROUGH 2016

As of May 2016, U.S. Immigration and Customs Enforcement (ICE) reported that there were about 950,100 aliens with final orders of removal on ICE's docket, of which about 182,200, or 19 percent, were criminals.[137] ICE's removals of aliens include individuals apprehended by ICE officers and individuals encountered by U.S. Customs and Border Protection agents and officers that are transferred to ICE for removal from the United States. According to ICE data, between 53 and 59 percent of all aliens ICE removed each year from fiscal years 2011 through 2016 were criminal aliens.[138] The total number of criminal aliens ICE removed from the United

---

[137] ICE defines a criminal alien as an alien convicted of a crime, either within or outside of the United States. According to ICE, these figures include individuals who could not lawfully be removed due to certain protections afforded under the Immigration and Nationality Act, such as temporary protective status or withholding of removal; individuals who may be lawfully removed but who are no longer enforcement priorities; individuals who are enforcement priorities but who have been released under conditions (e.g., electronic monitoring, regular reporting requirements, bond) due to case-specific circumstances; and individuals who are enforcement priorities and are targeted for removal through ICE's increased at-large operations, such as fugitives with criminal convictions. These figures also include aliens whose removal ICE is coordinating and aliens whose departure ICE has been unable to confirm. Additionally, there are some countries that do not cooperate with the U.S. government in accepting the return of their citizens who have final removal orders from the United States. Thomas Homan, Enforcement and Removal Operations, U.S. Immigration and Customs Enforcement, response to questions for the record from the Senate Committee on the Judiciary, Subcommittee on Immigration and the National Interest regarding the hearing, *Declining Deportations and Increasing Criminal Alien Releases–The Lawless Immigration Policies of the Obama Administration*, 114th Cong., 2nd sess., May 19, 2016.

[138] These data are reported by ICE in its annual *ICE Enforcement and Removal Operations Report* and includes removals administered by ICE. ICE's removal data differs from the removal data reported by DHS's Office of Immigration Statistics in its *Yearbook of Immigration Statistics* and *Annual Report on Immigration Enforcement Actions* because the DHS reports include removals administered by U.S. Customs and Border Protection. In addition, according to DHS, the removal and return numbers in these reports are estimates, largely because U.S. Customs and Border Protection records indicate which aliens the agency intends to remove and do not have explicit records of its removals. DHS's *Annual Report on Immigration Enforcement Actions* reported approximately 340,100 total removals for fiscal year 2016. Of those, ICE removed approximately 228,200 individuals, U.S. Border Patrol removed approximately 85,000 individuals, and U.S. Customs and Border Protection Office of Field Operations removed approximately 26,100 individuals.

States decreased in recent years–from about 216,000 in fiscal year 2011 to about 139,000 in fiscal year 2016. As illustrated in Figure 30, the subset of these criminal aliens who were apprehended by ICE also decreased from fiscal years 2011 through 2016, from about 150,000 to about 60,000 removals. Of all aliens apprehended by ICE who were subsequently removed from the United States, the proportion of those aliens who were criminal aliens increased each year, from 67 percent in 2011 to 92 percent in 2016.

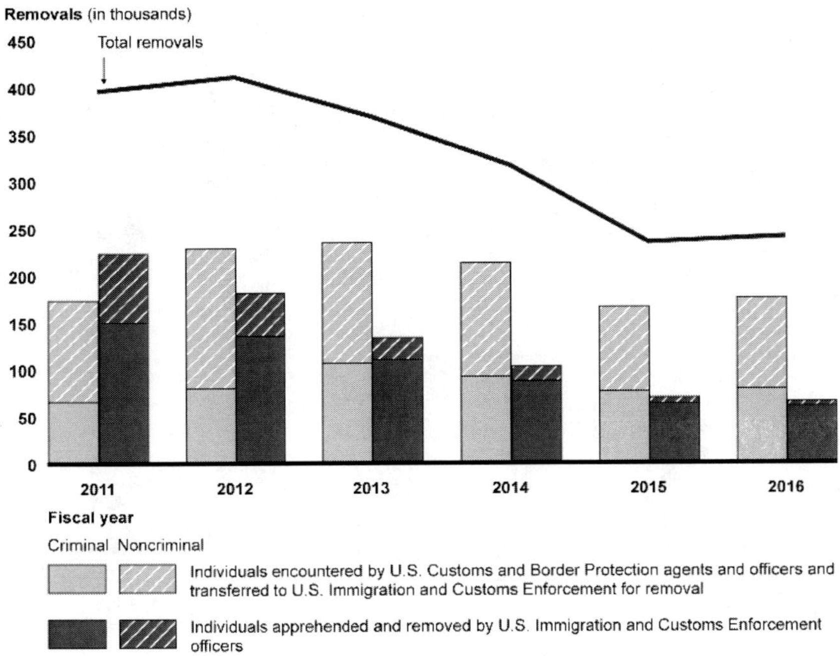

Source: GAO analysis of U.S. Immigration and Customs Enforcement data. | GAO-18-433.

Note: ICE defines a criminal alien as an alien convicted of a crime, either within or outside of the United States.

Figure 30. U.S. Immigration and Customs Enforcement's Removals from the United States, Fiscal Years 2011 through 2016.

Appendix X: Comments from the Department of Justice

U.S. Department of Justice

Washington, D.C. 20530

July 12, 2018

Gretta L. Goodwin, Ph.D.
Director, Justice and Law Enforcement Issues
Homeland Security and Justice Team
U.S. Government Accountability Office
441 G Street N.W.
Washington, DC 20548

Dear Dr. Goodwin:

Thank you for the opportunity to review and comment on your draft report titled "*Criminal Alien Statistics: Information on Incarcerations, Arrests, Convictions, Costs, and Removals* (GAO-18-433). In this report, the Government Accountability Office (GAO) presented a number of statistics regarding the apprehension, arrest, prosecution, and incarceration of aliens who committed crimes.

The Department of Justice (DOJ) appreciates the GAO's work to provide additional information on incarcerations, arrests, convictions, costs, and removals of criminal aliens. However, we explain, below, why some of the statistics cited could be misinterpreted or misunderstood by those who read the report.

Although the use of data from the State Criminal Alien Assistance Program (SCAAP) provides some insight into the level of criminality of aliens at the state and local level, as the draft report we reviewed indicates in a few locations, SCAAP information is generally not a reliable indicator of the total criminality levels of aliens at the state and local level. Our concern is that the draft report—while appropriately qualifying data throughout—does not highlight early enough in the report the limitations on using SCAAP information to avoid misinterpretation or misunderstanding by recipients of the report.

For one, the very definitional provisions of covered aliens employed in the SCAAP program leave out large numbers of criminal aliens. *See* 8 U.S.C. § 1231(i)(3)(B) (defining a covered alien as one who "has been convicted of a felony or two or more misdemeanors" and who either entered the United States unlawfully, was subject to ongoing immigration proceedings when taken into state or local custody, or who was admitted on a nonimmigrant visa but failed to maintain or comply with the conditions of their admission). Further, as GAO noted in the report, SCAAP data "does not include (a) aliens incarcerated in states or localities that did not apply for and receive federal reimbursement for costs of incarceration and (b) aliens with lawful immigration status who were not eligible for removal proceedings at the time they were taken into custody." U.S. Government Accountability Office, GAO-18-433, Criminal Alien Statistics: Information on Incarcerations, Arrests, Convictions, Costs, and Removals at 14 (2018). It is clear that this definition excludes potential large numbers of crimes committed by aliens at the state and local level—such as an alien who has been convicted of one serious misdemeanor such as Driving Under the Influence.

Similarly, citations to reduced costs associated with the program on an annual basis, or to declines in the total SCAAP population, do not accurately reflect the total criminality level of aliens at the state and local level. The draft report we reviewed contained appropriate qualifications—such as footnote 43 on page 14—but such qualifications would be more appropriately highlighted in a different manner to ensure that an individual reading the report does not misinterpret or misunderstand the data.

Conclusion

Thank you again for the opportunity to review and comment on this report. We look forward to working with the GAO and continuing to build on our successful relationship. Furthermore, we encourage the GAO to make the effort to explain accurately to others the inferences that might be drawn from this report regarding the numbers of criminal aliens in the law enforcement system.

Sincerely,

Michael H. Allen

for Lee J. Lofthus
Assistant Attorney general
for Administration

In: Top 5 Reports and Testimonies … ISBN: 978-1-53615-511-2
Editors: C. Johnston and W. Payne

Chapter 4

Critical Infrastructure Protection: DHS Should Take Actions to Measure Reduction in Chemical Facility Vulnerability and Share Information with First Responders*

United States Government Accountability Office

Abbreviations

CAMEO	Computer-Aided Management of Emergency Operations
CFATS	Chemical Facility Anti-Terrorism Standards
CSAT	Chemical Security Assessment Tool
DHS	Department of Homeland Security
EPCRA	Emergency Planning and Community

* This is an edited, reformatted and augmented version of the United States Government Accountability Office Report to Congressional Committees, Publication No. GAO-18-538, dated August 2018.

	Right-to-Know Act of 1986
GPRA	Government Performance and Results Act
HSSAI	Homeland Security Studies and Analysis Institute
IP Gateway	Infrastructure Protection Gateway
ISCD	Infrastructure Security Compliance Division
LEPC	Local Emergency Planning Committee
NIPP	National Infrastructure Protection Plan
SERC	State Emergency Response Commission
TEPC	Tribal Emergency Planning Committee
TERC	Tribal Emergency Response Commission

WHY GAO DID THIS STUDY

Facilities that produce, use, or store hazardous chemicals could be targeted or used by terrorists to inflict mass casualties, damage, and fear. DHS established the CFATS program to assess the risk posed by these facilities and inspect them to ensure compliance with DHS standards. DHS places high-risk facilities in risk-based tiers and is to conduct inspections after it approves their security plans. Under the CFATS Act of 2014, authorization for the CFATS program expires in January 2019.

GAO assessed the extent to which DHS has (1) enhanced the process for identifying high-risk facilities and assigning them to tiers, (2) conducted facility inspections and measured facility security, and (3) ensured that information is shared with emergency responders to prepare them for incidents at high-risk facilities. GAO reviewed DHS reports and data on compliance inspections and interviewed DHS officials. GAO also obtained non-generalizable information from 11 trade associations representing chemical facilities regarding DHS outreach and from 15 emergency planning committees about their awareness of CFATS and the chemicals it covers.

What GAO Recommends

GAO recommends that DHS take actions to (1) measure reduction in vulnerability of high-risk facilities and use that data to assess program performance; and (2) encourage access to and wider use of the IP Gateway among first responders and emergency planners. DHS concurred with both recommendations and outlined efforts underway or planned.

What GAO Found

Since 2013, the Department of Homeland Security (DHS) has strengthened its processes for identifying high-risk chemical facilities and assigning them to tiers under its Chemical Facility Anti-Terrorism Standards (CFATS) program. Among other things, DHS implemented a quality assurance review process to verify the accuracy of facility self-reported information used to identify high-risk facilities. DHS also revised its risk assessment methodology—used to assess whether chemical facilities are high-risk and, if so, assign them to a risk-based tier—by incorporating changes to address prior GAO recommendations and most of the findings of a DHS-commissioned peer review. For example, the updated methodology incorporates revisions to the threat, vulnerability, and consequence scoring methods to better cover the full range of security issues regulated by CFATS. As of February 2018, a total of 29,195 facilities—including all 26,828 facilities previously assessed and 2,367 facilities new to the program—were assessed using DHS's revised methodology. DHS designated 3,500 of these facilities as high-risk and subject to further requirements.

DHS has also made substantial progress conducting and completing compliance inspections and has begun to take action to measure facility security but does not evaluate vulnerability reduction resulting from the CFATS compliance inspection process. In 2013, GAO found that the backlog of chemical facility security plans awaiting review affected DHS's ability to conduct compliance inspections, which are performed after security plans are approved. Since then DHS has made progress and increased the number of completed compliance inspections. As of May 2018, DHS had conducted 3,553 compliance inspections. DHS has also begun to update its performance measure for the CFATS program to evaluate security measures implemented both when a facility submits its initial security plan and again when DHS approves its final security plan. However, GAO found that DHS's new performance measure methodology does not measure reduction in vulnerability at a facility resulting from the implementation and verification of planned security measures during the compliance inspection process. Doing so would provide DHS an opportunity to begin assessing how vulnerability is reduced—and by extension, risk lowered—not only for individual high-risk facilities but for the CFATS program as a whole.

DHS shares some CFATS information, but first responders and emergency planners may not have all of the information they need to minimize the risk of injury or death when responding to incidents at high-risk facilities. Facilities are currently required to report some chemical inventory information, but GAO found that over 200 CFATS chemicals may not be covered by these requirements. To improve access to information, DHS developed a secure interface called the Infrastructure Protection (IP) Gateway that provides access to CFATS facility-specific information that may be missing from required reporting. However, GAO found that the IP Gateway is not widely used at the local level. In addition, officials from 13 of the 15 Local Emergency Planning Committees—consisting of first responders and covering 373 CFATS high-risk facilities—told GAO they did not have access to CFATS data in the IP Gateway. By encouraging wider use of the IP Gateway, DHS would have greater assurance that first responders have information about high-risk facilities and the specific chemicals they possess.

August 8, 2018

The Honorable Ron Johnson
Chairman

The Honorable Claire McCaskill
Ranking Member
Committee on Homeland Security and Governmental Affairs
United States Senate

The Honorable Michael T. McCaul
Chairman

The Honorable Bennie G. Thompson
Ranking Member
Committee on Homeland Security
House of Representatives

Thousands of facilities that produce, use, or store hazardous chemicals could be targeted or used by terrorists to inflict mass casualties, damage, and fear. These chemicals could be released from a facility to cause harm to surrounding populations; they could be stolen and used as chemical weapons or as their precursors (the ingredients for making chemical weapons); or they could be stolen and used to build an improvised explosive device. Past incidents demonstrate the danger these chemicals pose, including the 2013 ammonium nitrate explosion at a fertilizer storage and distribution facility in West, Texas, which killed 15 people and caused major damage to or destroyed 193 homes, and more recent high-profile international incidents such as attacks using chlorine in Syria.

Pursuant to the Department of Homeland Security (DHS) Appropriations Act, 2007, DHS established the Chemical Facility Anti-Terrorism Standards (CFATS) program to, among other things, identify high-risk chemical facilities and assess the risk posed by each; place facilities identified as high-risk into one of four risk-based tiers; and assess, approve, and inspect facility security measures to ensure compliance with regulatory requirements.[1] The Protecting and Securing Chemical Facilities from Terrorist Attacks Act of 2014 (CFATS Act of 2014), enacted in December 2014, in effect, reauthorized the CFATS program for an additional 4 years while also imposing additional implementation requirements on DHS for the program.[2] DHS's National Protection and Programs Directorate's Infrastructure Security Compliance Division (ISCD) is responsible for managing the CFATS program.

We previously reported on various aspects of the CFATS program and identified challenges DHS was experiencing in implementing and managing the program. We made a number of recommendations to strengthen the program to include, among other things, that DHS verify that certain data reported by facilities are accurate, enhance its risk assessment approach to incorporate all elements of risk, conduct a peer review of the program to validate and verify DHS's risk assessment approach, document processes and procedures for managing compliance with site security plans, and update the performance measure for the program. DHS agreed with our recommendations and has either fully implemented them or taken action to begin addressing them.[3]

---

[1] See 72 Fed. Reg. 17,792 (Apr. 9, 2007) (codified as amended at 6 C.F.R. pt. 27); see also Pub. L. No. 109-295, § 550, 120 Stat. 1355, 1388-89 (2006).

[2] See Pub. L. No. 113-254, 128 Stat. 2898 (2014); 6 U.S.C. §§ 621-629. Specifically, the Act amended the Homeland Security Act of 2002, Pub. L. No. 107-296, 116 Stat. 2135 (2002), as amended, by adding Title XXI—Chemical Facility Anti-Terrorism Standards— and expressly repealing the program's authority under the fiscal year 2007 DHS appropriations act.

[3] GAO, *Critical Infrastructure Protection: DHS Is Taking Action to Better Manage Its Chemical Security Program, but It Is Too Early to Assess Results*, GAO-12-515T (Washington, D.C.: July 26, 2012); *Critical Infrastructure Protection: DHS Efforts to Assess Chemical Security Risk and Gather Feedback on Facility Outreach Can Be Strengthened*, GAO-13-353 (Washington, D.C.: Apr. 5, 2013); *Critical Infrastructure Protection: DHS Efforts to Identify, Prioritize, Assess, and Inspect Chemical Facilities*, GAO-14-365T (Washington, D.C.: Feb. 27, 2014); *Critical Infrastructure Protection: Observations on DHS Efforts to Implement and Manage Its Chemical Security Program*, GAO-14-608T (Washington, D.C.: May 14, 2014); *Critical Infrastructure Protection: DHS Action Needed to Verify Some Chemical Facility Information and Manage Compliance Process*, GAO-15-614 (Washington, D.C., July 22, 2015); *Critical Infrastructure Protection: Improvements Needed for DHS's Chemical Facility Whistleblower Report Process*, GAO-16-572 (Washington, D.C.: Jul 12, 2016); and *Critical Infrastructure Protection: DHS Has Fully Implemented Its Chemical Security Expedited Approval Program and Participation To Date Has Been Limited*, GAO-17-502 (Washington, D.C.: June 29, 2017).

Although there have been program improvements in recent years, questions remain about the progress DHS has made implementing changes to the program and the extent to which CFATS is ensuring that the highest-risk chemical facilities are more secure as a result. Given that the authorization for the CFATS program expires in January 2019, you requested that we assess the progress DHS has made implementing and managing the CFATS program—both within the context of our prior work and with regard to other areas—to inform legislative efforts related to reauthorization of the program. This chapter discusses the extent to which the CFATS program has taken action to (1) enhance the process for identifying high-risk chemical facilities and assigning them to risk-based tiers, (2) conduct chemical facility inspections and implement an approach to measure facility security, and (3) ensure that information is shared with first responders and emergency planners to prepare them for incidents at high-risk chemical facilities.

To address our first objective, we reviewed our prior work and analyzed documentation and data describing ISCD's efforts to assess, update, and implement the program's risk assessment methodology since we first evaluated it in fiscal year 2013. Specifically, we analyzed ISCD documents describing the web-based tools used to collect security information from facilities, and policies and procedures for reviewing and validating the accuracy of this information. We also reviewed relevant technical reports, plans, and assessments—including the findings and recommendations of an ISCD-commissioned peer review and DHS actions taken in response—describing changes made to ISCD's risk assessment methodology. In addition, we obtained data describing the status of ISCD's efforts to reassess chemical facilities identified as high-risk and assign them a risk-based tier using the revised risk assessment methodology. We assessed the reliability of ISCD data by reviewing relevant documentation and interviewing knowledgeable officials about system controls. We concluded that these data were sufficiently reliable for the purposes of this chapter. We also interviewed ISCD officials to confirm our understanding of the documents and data provided, and actions ISCD has taken to enhance the process for identifying high-risk chemical facilities and assigning them risk-based tiers.

To address our second objective, we focused on actions DHS has taken to ensure compliance with the CFATS regulation since we first examined this area in 2015. We reviewed laws and regulations applicable to how DHS is to ensure compliance with the CFATS regulation and analyzed ISCD documents and data on the implementation status of the program's compliance inspection process. To examine the program's compliance inspection process, we analyzed procedures and guidance—such as ISCD's Standard Operating Procedure and Inspection Handbook for CFATS facility inspections, compliance inspection training documents, and CFATS Risk-Based Performance Standards Guidance. We also obtained data on the numbers of completed compliance inspections per year and the extent to which these inspections resulted in a corrective action. We assessed the reliability of ISCD compliance inspection data by reviewing relevant documentation and interviewing knowledgeable officials and concluded that these data were sufficiently reliable for the purposes of this chapter. We also interviewed ISCD officials to confirm our understanding of the changes made to the inspection process since fiscal year 2015. In addition, we conducted two site visits to observe scheduled compliance inspections at facility locations in Delaware and Maryland. While information obtained from these inspections cannot be generalized to all inspections, it provides insight and context on how ISCD officials implement compliance inspection procedures and guidance. Lastly, we contacted officials representing 15 trade associations with members that include a wide range of CFATS-regulated chemical facilities and who participated in the Chemical Sector Coordinating Council.[4] For the 11 trade associations that

[4] We selected these 15 trade associations because they are listed in the National Infrastructure Protection Plan (NIPP) as trade associations representing a high percentage of the Nation's Chemical Sector. According to the NIPP, working with these trade associations is a more manageable number of contact points through which DHS can coordinate activities with a large number of the asset owners and operators in the chemical sector. According to the NIPP, a Sector Coordinating Council is the principal entity under which owners and operators of critical infrastructure can coordinate with the government on a wide range of protection activities and issues. The Chemical Sector Coordinating Council represents owners and operators of chemical facilities. See DHS, *2013 National Infrastructure*

responded, we conducted semistructured interviews to obtain their perspectives on DHS's actions to communicate lessons learned to CFATS facilities on methods to reduce risk and the compliance inspection process. The information obtained from these 11 trade associations is not generalizable to the universe of chemical facilities covered by CFATS; however, it does provide insights into DHS's efforts to perform outreach and seek feedback on implementation of the CFATS program.

To determine actions taken to measure chemical facility security, we reviewed the CFATS regulation and analyzed ISCD reports and information describing efforts to update the performance measure for the CFATS program since we first evaluated it in fiscal year 2015. We interviewed ISCD officials to confirm our understanding of the changes made to revise the methodology of the performance measure and compared this new approach with criteria in the *National Infrastructure Protection Plan* (NIPP) for evaluating the effectiveness of risk management efforts by, among other things, collecting performance data to assess progress in achieving identified outputs and outcomes.[5]

To address our third objective, we reviewed laws, regulations, and other authorities applicable to how and to what extent DHS is to share CFATS data with first responders and emergency planners. We reviewed DHS documentation and interviewed officials to confirm our understanding of the types of information, methods used, and extent to which the program shares CFATS-specific information with first responders and emergency planners and compared DHS's approach with criteria in the NIPP on how agencies should share actionable and relevant information across the critical infrastructure community to build awareness and enable risk-informed decision making. To determine what information first responders and emergency planners may use to prepare for and

---

Protection Plan (NIPP), Partnering for Critical Infrastructure Security and Resilience (Washington, D.C.: December 2013).

[5] The NIPP risk management framework is a planning methodology that outlines the processes for, among other things, setting goals and objectives; identifying critical infrastructure; assessing risk based on consequences, threats, and vulnerabilities; implementing protective programs and resiliency strategies; and measuring performance and taking corrective actions. Broadly defined, risk management is a process that helps policymakers assess risk, strategically allocate finite resources, and take actions under conditions of uncertainty.

respond to emergencies at chemical facilities and the extent to which they are aware of CFATS facilities in their jurisdictions, we conducted semistructured interviews with officials representing a nonrandom sample of 15 Local Emergency Planning Committees (LEPCs).[6] We selected our sample of LEPCs—whose jurisdictions include 373 high-risk chemical facilities regulated by the CFATS program—to represent a range in the geographic location and numbers of CFATS facilities covered by each.[7] The information obtained from these interviews is not generalizable nor reflects the opinions of all first responders and emergency planners; however, it does provide insights into common themes and illustrative examples across our sample on the topics and issues discussed.

We also analyzed the list of chemicals, quantities, and concentrations regulated by the CFATS program and the chemical inventory reporting requirements outlined in the Emergency Planning and Community Right-to-Know Act of 1986 (EPCRA) to determine the extent to which there may be differences in the chemicals covered and reported by facilities subject to CFATS and EPCRA requirements.[8] Using the results of this analysis, we selected a generalizable random sample of 347 high-risk CFATS facilities and analyzed ISCD data on their chemical holdings to determine the extent to which there may be differences in the chemicals and quantities covered by CFATS and EPCRA and what facilities may be

---

[6] In accordance with the Emergency Planning and Community Right-to-Know Act of 1986 (EPCRA), state and local entities, such as Local Emergency Planning Committees (LEPCs), consisting of representatives including local officials and planners, facility owners and operators, first responders, and health and hospital personnel, among others, were established. See 42 U.S.C. § 11001. Among other things, LEPCs help communities prepare for and mitigate the effects of a chemical incident and ensure that information on chemical risks in the community is provided or is otherwise made available to emergency responders and the public.

[7] Our nonrandom sample was selected using an ISCD list of 3,516 CFATS facilities designated as high-risk (i.e., assigned to tiers 1 through 4) as of October 31, 2017. We selected LEPCs from different states to include counties from among those with the highest number of high-risk CFATS facilities in each state. The number of high-risk CFATS facilities located in each LEPC ranges from a low of 11 to a high of 88 across our sample.

[8] Under Section 312 of EPCRA, facilities are required to submit an emergency and hazardous chemical inventory form—referred to as a Tier II form. See 42 U.S.C. § 11022. The purpose of this form is to provide state and local officials and the public with specific information on potential hazards. This information includes the locations and amount of hazardous chemicals present at a facility during the previous calendar year.

required to report.[9] We assessed the reliability of ISCD data by reviewing relevant documentation and interviewing knowledgeable officials and concluded that these data were sufficiently reliable for the purposes of this chapter.

We conducted this performance audit from July 2017 to August 2018 in accordance with generally accepted government auditing standards. Those standards require that we plan and perform the audit to obtain sufficient, appropriate evidence to provide a reasonable basis for our findings and conclusions based on our audit objectives. We believe that the evidence obtained provides a reasonable basis for our findings and conclusions based on our audit objectives.

BACKGROUND

DHS's National Protection and Programs Directorate leads the country's effort to protect and enhance the resilience of the nation's physical and cyber infrastructure. The directorate includes the Office of Infrastructure Protection, which leads the coordinated national effort to reduce risk to U.S. critical infrastructure posed by acts of terrorism. Within the Office of Infrastructure Protection, ISCD leads the nation's effort to secure high-risk chemical facilities and prevent the use of certain chemicals in a terrorist act on the homeland; ISCD also is responsible for implementing and managing the CFATS program.

The CFATS program is intended to ensure the security of the nation's chemical infrastructure by identifying high-risk chemical facilities,

[9] Our generalizable random sample was selected using an ISCD list of 3,539 CFATS facilities designated as high-risk (i.e., assigned to tiers 1 through 4) as of January 8, 2018. The random sample of 347 facilities is generalizable with an expected margin of error no larger than plus or minus 5 percentage points at the 95 percent level of confidence. Because we followed a probability procedure based on random selections, our sample is only one of a large number of samples that we might have drawn. Since each sample could have provided different estimates, we express our confidence in the precision of our particular sample's results as a 95 percent confidence level margin of error (for example, plus or minus 5 percentage points). This forms the confidence interval that would contain the actual population value for 95 percent of the samples we could have drawn. Margins of error at the 95 percent level of confidence are provided along with each sample estimate in the report.

assessing the risk posed by them, and requiring the implementation of measures to protect them. Section 550 of the DHS Appropriations Act, 2007, required DHS to issue regulations establishing risk-based performance standards for chemical facilities that, as determined by DHS, present high levels of risk, to include vulnerability assessments and the development and implementation of site security plans for such facilities.[10] DHS published the CFATS interim final rule in April 2007 and Appendix A to the rule, published in November 2007, lists 322 chemicals of interest and the screening threshold quantities for each.[11] According to DHS, subject to certain statutory exclusions, all facilities that manufacture, store, ship, or otherwise use chemicals of interest above certain threshold quantities and concentrations are subject to CFATS reporting requirements.[12] However, only those facilities subsequently determined to present a high level of security risk are subject to the more substantive requirements of the CFATS regulation as described below.[13]

The CFATS Regulation and Process

The CFATS regulation outlines a specific process for how ISCD is to administer the CFATS program. A chemical facility that possesses any of 322 chemicals of interest in quantities that meet or exceed a threshold

---

[10] Pub. L. No. 109-295, § 550, 120 Stat. at 1388-89 (2006).

[11] 72 Fed. Reg. 17,688 (Apr. 9, 2007) (codified as amended at 6 C.F.R. pt. 27); 72 Fed. Reg. 65,396 (Nov. 20, 2007) (codified at 6 C.F.R. pt. 27, App. A). As of July 2018, the interim final rule (i.e., the CFATS regulation), as subsequently amended, remains in effect. Appendix A has not been revised since its initial publication.

[12] Such facilities can include food-manufacturing facilities that use chemicals of interest in the manufacturing process, universities that use the chemicals to do experiments, or warehouses that store ammonium nitrate, among others. Under the CFATS Act of 2014, such a facility may be recognized as a "chemical facility of interest." See 6 U.S.C. § 621(2). Consistent with law and regulation, certain facilities—including, in general, facilities regulated under the Maritime Transportation Security Act of 2002 (Public Law 107-295, 116 Stat. 2064), public water systems or wastewater treatment facilities, facilities owned and operated by the Department of Defense or the Department of Energy, and facilities subject to regulation by the Nuclear Regulatory Commission or in accordance with the Atomic Energy Act of 1954—are not subject to regulation under CFATS and are referred to as excluded facilities. 6 U.S.C. § 621(4).

[13] See generally 6 C.F.R. pt. 27, subpt. B.

quantity and concentration is required to complete what is called a Top-Screen survey using ISCD's Chemical Security Assessment Tool (CSAT) system. CSAT is a web-based application through which owners and operators of chemical facilities provide self-reported information about the facility. The Top-Screen is an on-line survey whereby the facility is to provide DHS various data, including the name and location of the facility and the chemicals, quantities, and storage conditions at the site.

ISCD uses a risk-based approach to evaluate chemical facilities of interest that are required to report under CFATS and determine whether these facilities are high-risk and therefore subject to further requirements under the regulation. More specifically, ISCD's risk assessment methodology calculates risk scores—based on facility-supplied information in the Top-Screen survey, among other sources, and taking into account vulnerability, potential consequences, and threat of a terrorist attack—and uses these scores to determine which facilities are high-risk. Those facilities deemed high-risk are then placed into one of four risk-based tiers (Tier 1 through Tier 4).[14] Tier 1 represents the highest risk. A facility not designated as high-risk is not subject to additional requirements under the CFATS regulation.[15]

If ISCD determines that a facility is high-risk (Tier 1–4), the facility must then complete and submit to ISCD a Security Vulnerability Assessment and one of two types of security plans—a Site Security Plan or an Alternative Security Program—which describes the existing and planned security measures to be implemented in order to be in compliance with the applicable risk-based performance standards.[16] Facilities

---

[14] For purposes of characterizing security risk, ISCD's risk assessment methodology is based on a range of potential attack scenarios generally organized across three security issues depending on the type of risk associated with the chemical of interest: (1) release (toxic, flammable, and explosive) chemicals with the potential for impacts within and beyond a facility; (2) theft or diversion; and (3) sabotage/contamination.

[15] A change in chemical holdings would require the facility to be reassessed using updated Top-Screen information and, where appropriate, assigned a risk tier.

[16] See 6 C.F.R. §§ 27.215 (security vulnerability assessments), 27.225 (site security plans), 27.235 (alternative security programs). DHS's CFATS regulation establishes 18 risk-based performance standards that identify the areas for which a facility's security posture are to be examined, such as perimeter security, access control, and cybersecurity. See 6 C.F.R. § 27.230.

determined to be Tier 3 or 4 also have an option to submit an expedited security plan under the CFATS Expedited Approval Program.[17] To meet risk-based performance standards, covered facilities may choose the security programs or processes they deem appropriate so long as ISCD determines that the facilities achieve the requisite level of performance on each of the applicable areas in their existing and agreed-upon planned measures. Prior to approving a facility's security plan, ISCD inspectors conduct an authorization inspection at the facility to verify and validate that the content listed in their plan is accurate and complete; that existing and planned equipment, processes, and procedures are appropriate and sufficient to meet the established requirements of the risk-based performance standards; and to assist the facility in resolving any potential gaps identified. After the facility's security plan is approved, the facility enters into the CFATS compliance cycle, which includes regular and recurring compliance inspections.

ISCD inspectors conduct compliance inspections to ensure the existing and planned security measures identified in a facility's approved security plan continue to be implemented fully; the equipment, processes, and procedures described in the security plan are appropriate and sufficient to meet the established performance standards; and the required corrective actions have been implemented and are sustainable. This compliance inspection includes a verification of other data provided to ISCD, including the Top-Screen. If, through a compliance inspection, ISCD determines a facility has not fully implemented security measures as outlined in its approved security plan, ISCD is to provide the facility with written notification that clearly identifies the deficiencies in the plan and will work with the facility toward achieving full compliance or, if warranted, take enforcement action. Figure 1 illustrates the CFATS regulatory process.

[17] The CFATS Expedited Approval Program was implemented in June 2015. DHS's expedited program guidance identifies specific security measures that eligible (i.e., Tiers 3 and 4) high-risk facilities can use to develop expedited security plans, rather than developing standard (non-expedited) security plans which provide more flexibility in securing a facility but are also more time-consuming to process. For more information, see GAO-17-502.

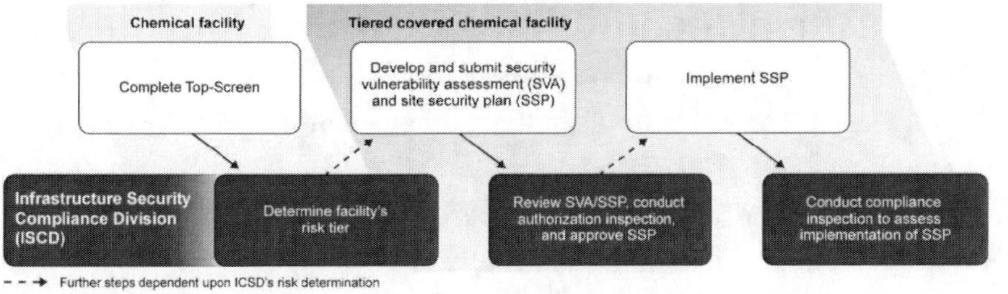

Source: GAO analysis of CFATS regulatory process. | GAO-18-538.

Figure 1. Department of Homeland Security's Chemical Facility Anti-Terrorism Standards (CFATS) Regulatory Process.

ISCD HAS STRENGTHENED ITS PROCESSES FOR IDENTIFYING HIGH-RISK CHEMICAL FACILITIES

ISCD Implemented Processes to Verify Self-Reported Information from Chemical Facilities

In response to our prior recommendations, ISCD has taken action to strengthen its processes for verifying the accuracy of data it uses to identify high-risk chemical facilities. In July 2015, we found that ISCD used self-reported and unverified data to determine the risk categorization for facilities that held toxic chemicals that could threaten surrounding communities if released.[18] At the time, ISCD required that facilities self-report the Distance of Concern—an area in which exposure to a toxic chemical cloud could cause serious injury or fatalities from short-term exposure—as part of its Top-Screen methodology. In our report, we estimated that more than 2,700 facilities with a toxic release threat misreported the Distance of Concern and recommended that ISCD (1) develop a plan to implement a new Top-Screen to address errors in the Distance of Concern submitted by facilities, and (2) identify potentially

[18] GAO-15-614.

miscategorized facilities that could cause the greatest harm and verify that the Distance of Concern these facilities reported is accurate.[19]

ISCD has addressed both of these recommendations. In response to the first recommendation, ISCD implemented an updated Top-Screen survey in October 2016 and now collects data from facilities and conducts more accurate modeling to determine the actual area of impact (formerly called the Distance of Concern), rather than relying on the facilities' calculation. In response to the second recommendation, ISCD officials reported in November 2016 that they reassessed all facility Top-Screens that reported threshold quantities of chemicals posing a toxic release threat, and identified 158 facilities with the potential to cause the greatest harm. In April 2018, ISCD officials reported that all of these facilities have since been reassessed using updated Top-Screen information and, where appropriate, assigned a risk tier.

In addition, in October 2016, ISCD implemented a quality assurance review process whereby ISCD officials manually check and verify the accuracy of facility self-reported Top-Screen information used in identifying potential high-risk facilities. The objective of ISCD's review process is to evaluate the information provided by a chemical facility in order to recommend approval or rejection of a submitted Top-Screen for accuracy prior to issuing a letter notifying the facility of its risk tier designation. According to ISCD, all Top-Screens undergo a quality assurance review with two exceptions: (1) a facility that registers through CSAT for the first time and submits a Top-Screen identifying zero chemicals of interest on site and which does not identify an exclusion; or (2) a facility that possessed a chemical of interest in the past but subsequently submits a follow-up Top-Screen for redetermination identifying they no longer possess the chemical of interest and after ISCD validates the removal of the chemical of interest. When a Top-Screen

[19] Specifically, we recalculated the Distance of Concern for a generalizable sample of facilities—a simple random sample of 475 facilities from the population of 36,811 facilities that submitted Top-Screens since the inception of the CFATS program in 2007 through January 2, 2015—and compared these results to what facilities reported in their Top-Screen submission. Based upon this sample, we estimated that 4,173 facilities with a toxic release chemical misreported the Distance of Concern, with an associated 95 percent confidence interval of 2,798 to 5,822 facilities.

submission is rejected, ISCD sends a letter notifying the facility of the rejection and requesting that a revised Top-Screen be submitted. In addition, according to ISCD, they contact facilities prior to a Top-Screen rejection to ensure the facility understands the required updates and to discuss the potential reporting error. As of February 2018, a total of 1,956 Top-Screen submissions (across 1,799 unique facilities) were rejected as part of this quality assurance review process since implementing the updated Top-Screen survey in October 2016, according to ISCD data. According to ISCD, the majority of these Top-Screens were rejected due to common reporting errors, such as misreporting the flammability hazard rating for a chemical of interest subject to a release security issue or not reporting transportation packaging when a chemical of interest is identified as being subject to a theft or diversion security issue.

ISCD Has Nearly Completed Applying Its Revised Risk Assessment Methodology for Designating High-Risk Chemical Facilities

ISCD Revised Its Risk Assessment Methodology to More Accurately Identify and Assign Tiers to High-Risk Chemical Facilities

Since we last evaluated it in 2013, ISCD took action to enhance the CFATS program's risk assessment methodology—used to determine whether covered chemical facilities are high-risk and, if so, assign them a risk-based tier—by incorporating changes to address prior GAO recommendations, as well as the findings of an ISCD-commissioned peer review conducted in 2013, among other efforts. In April 2013, we found that DHS's risk assessment approach did not consider all of the elements of threat, vulnerability, and consequence associated with a terrorist attack involving certain chemicals.[20] Our work showed that DHS's CFATS risk assessment methodology was based primarily on consequences from human casualties, but did not consider economic consequences, as called

[20] GAO-13-353.

for by the NIPP and the CFATS regulation. We also found that DHS's approach was not consistent with the NIPP because it treated every facility as equally vulnerable to a terrorist attack regardless of location or on-site security. In addition, DHS was not using threat data for 90 percent of the tiered facilities—those tiered for the risk of theft or diversion—and using 5-year-old threat data for the remaining 10 percent of those facilities that were tiered for the risks of release or sabotage.[21] We recommended that ISCD (1) review and improve its risk assessment approach to fully address each of the elements of threat, vulnerability, and consequence, and (2) conduct an independent peer review after enhancements to the risk assessment approach were complete.

Partly in response to our findings and recommendations, from 2013 through 2016, ISCD conducted a multiyear effort to review and improve the CFATS program's risk assessment approach and tiering methodology with the primary goal of improving the identification and appropriate tiering of high-risk chemical facilities. Among these efforts was an ISCD-commissioned peer review of the CFATS tiering methodology conducted in 2013 by the Homeland Security Studies and Analysis Institute (HSSAI).[22] HSSAI's final report summarized the findings of the peer review and included a list of 44 recommendations for ISCD to implement in its efforts to improve and revise the CFATS risk assessment and tiering methodology.[23] ISCD undertook a risk assessment improvement project to implement most of the recommendations described in the 2013 HSSAI final report; these efforts included, for example, convening advisory board

---

[21] For theft or diversion, DHS's model assumes that a terrorist will steal or have the chemical of interest diverted to him or herself and then estimates the risk of a terrorist attack using the chemical of interest in a way that causes the most harm at an unspecified off-site location.

[22] The Homeland Security Act of 2002 authorizes the Secretary of Homeland Security, acting through the Under Secretary for Science and Technology, to establish one or more federally funded research and development centers—such as the Homeland Security Studies and Analysis Institute (HSSAI)—to provide independent analysis of homeland security issues. See Pub. L. No. 107-296, § 305, 116 Stat. 2135, 2168 (2002); 6 U.S.C. § 185. HSSAI provides the government with, among other things, strategic studies and assessments. HSSAI also works with and supports other federal, state, local, tribal, public and private sector organizations.

[23] Mukta Agrawal, et al., *Chemical Facility Anti-Terrorism Standards Tiering Methodology Peer Review: Final Report* (Falls Church, Va: Homeland Security Studies and Analysis Institute, 2013).

meetings with experts drawn from across industry, academia, and government to review and make additional recommendations on the proposed improvements to the CFATS risk assessment methodology and associated tools and processes.

The result of these efforts is an updated, "second generation" risk assessment approach and tiering methodology that addresses both of our prior recommendations and almost all of the recommendations described in the 2013 HSSAI final report. Specifically, with regard to our recommendation that DHS enhance its risk assessment approach to incorporate all elements of risk, ISCD worked with Sandia National Laboratories to develop and evaluate a model to estimate the economic consequences of a chemical attack. In addition, among other enhancements, the updated risk assessment methodology incorporates revisions to the threat, vulnerability, and consequence scoring methods to better cover the full range of chemical security issues regulated by the CFATS program.[24] Additionally, with regard to our recommendation that DHS conduct a peer review after enhancing its risk assessment approach, DHS conducted peer reviews and technical reviews with government organizations and facility owners and operators, and worked with Sandia National Laboratories to verify and validate the CFATS program's revised risk assessment methodology which was completed in January 2017. In addition, as of May 2018, ISCD has considered, implemented, or is in the process of implementing updates that address 39 of the 44 recommendations in the HSSAI peer review of the original CFATS risk assessment methodology. According to ISCD, DHS must undertake a rulemaking to update the CFATS regulation and to obtain public comment on any proposed changes to implement the remaining recommendations. These relate to possible changes in how or to what extent the CFATS

---

[24] Under the CFATS program, chemicals of interest (chemicals of interest) are organized by security and vulnerability issues – specifically, chemicals of interest that a terrorist could: steal, divert, or otherwise acquire to use as a weapon at another time and place (theft/diversion chemicals of interest); or sabotage or contaminate to explode or release in transit (sabotage chemicals of interest); or release as an explosive or to form a flammable or toxic cloud (release chemicals of interest).

program regulates the treatment of certain chemicals of interest, chemical weapons and their precursors, and other fuels or fuel mixtures.[25]

Implementation of the Revised Risk Assessment Methodology Is Nearly Complete

Beginning in October 2016, ISCD notified chemical facilities that were not new to the CFATS program—that is, all facilities that had previously submitted a Top-Screen and had reported chemicals of interest above the threshold quantity and concentration on their most recent Top-Screen—to submit a revised Top-Screen in CSAT 2.0 so that they may be reassessed using ISCD's revised risk assessment methodology.[26] As of February 2018, a total of 29,195 chemical facilities were assessed using ISCD's revised risk assessment methodology, with 3,500 (or 12 percent) of these facilities designated as high-risk (i.e., assigned to tiers 1 through 4). The total of 29,195 chemical facilities includes 26,828 facilities that were previously assessed using the original risk assessment methodology and an additional 2,367 facilities new to the CFATS program, as shown in figure 2.[27]

---

[25] For example, DHS issued an Advance Notice of Proposed Rulemaking in August 2014 in an effort to more fully mature the CFATS program by helping the department identify how to make the CFATS program more effective in achieving its regulatory objectives. 79 Fed. Reg. 48,693 (Aug. 18, 2018). In particular, DHS expressed interest in comments addressing the general regulatory approach, treatment of non-traditional chemical facilities, clarification of terminology, Risk Based Performance standards, Appendix A, consideration for small businesses, and alignment with other regulatory programs. Id. at 48,694. The comment period for this Advance Notice ended on October 17, 2014. As of August 2017, DHS was in the process of reviewing public comments received and anticipated issuing a notice of proposed rulemaking in October 2017. 82 Fed. Reg. 40290, 40,292 (Aug. 24, 2017). As of January 2018, however, DHS remained in the process of reviewing the public comments received and the anticipated issuance of the notice of proposed rulemaking was "To Be Determined." 83 Fed. Reg. 1872, 1874 (Jan. 12, 2018). As of July 2018, DHS had not issued the anticipated notice of proposed rulemaking.

[26] In fall 2016, DHS revised its Chemical Security Assessment Tool (CSAT), which supports DHS efforts to gather information from facilities to assess their risk using the online Top-Screen survey. The new tool—called CSAT 2.0—eliminates duplication and confusion associated with DHS's original CSAT. DHS officials report that they improved the tool by revising some questions in the original CSAT to make them easier to understand; eliminating some questions; and prepopulating data from one part of the tool to another so that users do not have to retype the same information multiple times.

[27] In 2015 we reported that, since 2007, when ISCD began identifying chemical facilities to determine which facilities present a high risk and therefore should be subject to further regulation under CFATS, about 37,000 facilities had submitted a Top-Screen. ISCD officials acknowledged some facilities may have failed to do so but that they believed the

Original tiering	Numbers of unique chemical facilities	Breakdown of new tiering results: Tier 1	Tier 2	Tier 3	Tier 4	Not tiered (i.e. not high-risk)	Total
Tier 1	107 →	44	7	32	21	3	107
Tier 2	338 →	33	17	100	153	35	338
Tier 3	817 →	42	17	354	315	89	817
Tier 4	1,514 →	24	15	586	586	303	1,514
Previously not tiered	24,052 →	14	20	228	627	23,163	24,052
Total	26,828	157	76	1,300	1,702	23,593	26,828
Chemical facilities new to CFATS program		4	3	105	153	2,102	2,367
Total		161	79	1,405	1,855	25,695	29,195
		3,500 (Tiers 1–4)					

Source: GO analysis of CFATS data. | GAO-18-538.

Figure 2. Implementation Results of Chemical Facility Anti-Terrorism Standards (CFATS) Revised Risk Assessment Methodology (as of February 2018).

Of the 3,500 tiered facilities, 265 were new to the CFATS program; 889 were not new to the program, but were previously not tiered and were reassessed as high-risk and assigned a tier; and 1,345 were previously tiered but were reassigned to a different tier. Also, 430 facilities that were previously tiered were no longer tiered. As of May 2018, ISCD had pending risk assessments for an additional 241 chemical facilities that were not new to the CFATS program but were not previously tiered. ISCD officials did not provide an estimated target completion date for these pending risk assessments, noting that completing the assessments is highly dependent on the facilities providing the necessary Top-Screen information.

According to ISCD, there are four main drivers of the changes in facility tiering that resulted from implementing the second-generation risk assessment methodology:

37,000 facilities represent most facilities subject to CFATS. According to an ISCD official, as of June 2018, just over 40,000 facilities have submitted a Top-Screen.

- facilities placed in a lower tier due to implementation of revised consequence scoring methods that more accurately account for the impact of quantities of the chemicals subject to theft/diversion security issues;
- facilities placed in a higher or lower tier for chemicals of interest due to improvements to the distribution of population in consequence modeling for chemicals subject to release-toxic and release-flammable security issues;
- increases in the number of facilities tiered for select chemical weapon precursors due to the implementation of revised consequence scoring methods that more accurately account for the impact of certain chemicals of interest; and
- changes in tiering due to newly reported increases, decreases, and modifications of chemical holdings.

ISCD Has Made Progress Conducting Compliance Inspections but Does Not Measure Reduction in Facility Vulnerability

ISCD Has Increased the Number of Completed Compliance Inspections and Issued Two Corrective Actions for Noncompliance with Security Plans

Since 2013, ISCD has reduced its backlog of unapproved site security plans and increased the number of conducted compliance inspections. As discussed earlier, in order to approve a facility's site security plan, ISCD inspectors conduct an authorization inspection at the facility to verify and validate that the content listed in their plan is accurate and complete; that existing and planned equipment, processes, and procedures are appropriate and sufficient to meet the established requirements of the risk-based performance standards; and to assist the facility in resolving any potential gaps identified. After the facility's security plan is approved, the facility enters into the CFATS compliance cycle and is subject to a compliance inspection. In 2013, we calculated that it could take from 7 to 9 years to review and approve the approximately 3,120 site security plans submitted by facilities that had been designated as high-risk but that ISCD had not yet begun to review.[28] In 2015, we found that ISCD had made improvements to its processes for reviewing and approving site security plans and substantially reduced the time needed to approve remaining site plans to between 9 and 12 months.[29]

Our analysis of ISCD data since our 2015 report showed that ISCD has made substantial progress conducting and completing compliance inspections. Specifically, our analysis showed that ISCD has increased the number of compliance inspections completed per year since ISCD began conducting compliance inspections in 2013. For the 2,466 high-risk facilities with an approved site security plan as of May 2018, ISCD had conducted 3,553 compliance inspections.[30] Table 1 shows the number of conducted compliance inspections from fiscal year 2014 to May 2018.

Table 1. Number of Chemical Facility Anti-Terrorism Standards Compliance Inspections Conducted, by Fiscal Year (as of May 2018)

Fiscal year		Compliance inspections
2014		50
2015		155
2016		1,052
2017		1,571
2018	(October 2017-May 2018)	725
Total		3,553

Source: GAO analysis of Department of Homeland Security data. | GAO-18-538.

---

[28] GAO-13-353.

[29] .GAO-15-614.

[30] In accordance with the CFATS regulations, as a general matter, DHS intends to require facilities in Tiers 1 and 2 to update their Top-Screen every 2 years, and for Tiers 3 and 4 every 3 years. ISCD conducts compliance inspections on a regular and recurring basis. ISCD officials stated that compliance inspections are prioritized based on several factors including tier and the number of planned security enhancements required at facilities.

ISCD officials project they will conduct fewer compliance inspections in fiscal year 2018 than in fiscal year 2017 due to two reasons. First, ISCD officials stated the program made progress resolving the backlog of facilities that required compliance inspections in fiscal years 2016 and 2017 when it conducted over 2,600 compliance inspections. Second, ISCD officials stated that the program's revised risk assessment approach and continued outreach efforts have resulted in an increase in the number of identified facilities with chemicals of interest.[31] As a result, ISCD officials stated they project an increased number of authorization inspections and fewer compliance inspections in fiscal year 2018 and 2019 as new facilities enter the program.

ISCD increased the number of compliance inspections conducted from fiscal years 2014 to 2017 and less than 1 percent of compliance inspections during this period resulted in a determination that a facility was not in compliance. During a compliance inspection, if an inspector finds that a facility is noncompliant with its security plan, the CFATS regulation authorizes ISCD to take enforcement action, such as issuing an order for corrective action to the facility.[32] Of the 3,553 compliance inspections ISCD conducted between fiscal year 2014 and May 2018, ISCD issued two corrective actions—both to Tier 4 facilities—because these facilities were not in compliance with their approved site security plan.[33]

Specifically, during the compliance inspection of one facility, which was determined to be high-risk based on both the release and theft/diversion security issues, ISCD found that the facility's site security

---

[31] As part of DHS's plans to implement The Protecting and Securing Chemical Facilities from Terrorist Attacks Act of 2014, DHS established an outreach implementation plan in coordination with public and private stakeholders in order to, among other things, identify chemical facilities of interest that may be subject to regulations under CFATS. DHS updated its outreach implementation plan in fiscal year 2018.

[32] See 6 C.F.R. § 27.300. If through a compliance inspection it is determined a facility has not fully implemented security measures as outlined in its approved site security plan, ISCD is to provide the facility with written notification that clearly identifies the deficiencies in the site security plan and will work with the facility towards achieving full compliance or, if warranted, take enforcement action.

[33] In addition to these two corrective actions, since fiscal year 2015, DHS has issued five additional orders to four high-risk facilities with final penalties totaling $38,691.88. Of these five orders, three included the failure of a facility to submit an approvable security plan and two included the failure of a facility to submit a Top-Screen.

plan did not identify several existing or planned measures to secure the facility's chemicals of interest. For example, the facility's site security plan did not identify measures to monitor restricted areas or potentially critical targets within the facility against a theft or release attack. In addition, while the facility's site security plan identified a chain link fence and an alarm on a gate to a secure cage that houses the chemicals of interest, ISCD inspectors found no evidence of either. During the compliance inspection of the second facility, which was determined to be high-risk based on the theft and diversion security issue, ISCD inspectors were unable to verify if the facility's intrusion detection system was properly functioning and that an individual not employed by the facility may have had access to the facility's chemicals of interest without a proper background check. Both of these facilities took actions to implement the measures identified in their site security plan and were later found to be in compliance with their site security plans. ISCD officials attribute the low number of corrective actions the program has issued to the program's collaborative approach of working with facilities to ensure compliance. For example, of the two facilities ISCD found to be in noncompliance, ISCD conducted a compliance assistance visit with both facilities to provide assistance. In addition to compliance assistance visits, ISCD officials stated that the program has other collaborative tools, such as the CFATS Help Desk, to help ensure facility compliance.

ISCD Continues to Implement Changes to Compliance Inspections and Improve Efficiency

ISCD continues to implement changes that are intended to enhance compliance inspections. For example, ISCD officials stated the program continues to conduct preinspection phone calls with facilities to help them prepare for compliance inspections. In addition, ISCD officials stated they developed and provided supplemental guidance in fiscal year 2017 on steps ISCD inspectors need to take during a compliance inspection. ISCD's supplemental guidance includes, among other things, best practices and lessons learned for conducting inspections and reporting items identified by the inspections. ISCD officials stated they plan to incorporate this supplemental guidance into their compliance inspection standard operating procedures in the third quarter of fiscal year 2018 and to update their compliance inspection handbook in the fourth quarter of 2018.

In addition to updating its guidance for inspectors, ISCD has taken steps to improve the efficiency of compliance inspections. For example, ISCD continues its outreach efforts to chemical facilities on the inspection process. As part of these efforts, ISCD published guidance for facilities on steps to take to prepare for the compliance inspection, including information on the appropriate personnel and documentation that should be made available during the inspection. Finally, ISCD increased the number of compliance assistance visits with facilities to better prepare them for inspections.[34] Representatives from 9 of the 11 industry associations we spoke with told us that ISCD's communication with facilities had improved the efficiency of compliance inspections and increased the ability of facilities to comply with the risk-based performance standards.

We accompanied inspectors on two separate compliance inspections to observe how the inspections were carried out and how inspectors used the risk-based performance standards to determine compliance. For example, during the compliance inspection of a facility identified as high-risk based on the theft and diversion security issue, we observed facility personnel and ISCD inspectors discussing the preinspection phone call ISCD had conducted to assist the facility in preparation for their compliance inspection. This discussion included confirmation that the facility communicated with the local fire and police departments and had requested their presence at the inspection. In addition, we observed the inspectors analyzing the facility's emergency response plan to determine whether the facility's plans were consistent with the applicable risk-based performance

[34] DHS conducts compliance assistance visits to provide technical assistance and educate covered or potentially covered facilities on the CFATS regulation. Compliance assistance visits occur upon a facility's request or at DHS's request with the consent of the facility. During these visits, Inspectors can assist with submissions of Top-Screens, Security Vulnerability Assessments, or site security plans; assist facilities with registration; or answer additional questions, as necessary.

standards. We also observed the inspectors subsequently interviewing local fire and police department officials that were present during the inspection to validate statements made by the facility and to confirm that both entities received the facility's emergency plan. We accompanied the inspectors and facility personnel on a tour of the facility where inspectors observed existing measures the facility used to protect the chemicals of interest, including the facility's fencing barrier. We also observed inspectors testing security measures, including the facility's access controls put in place to prevent unauthorized personnel gaining access to the chemicals of interest.

At the other compliance inspection we observed, the facility personnel and ISCD inspectors confirmed a preinspection phone call to prepare the facility for the inspection. This phone call included a discussion of the appropriate training records and contract documentation that inspectors needed to confirm compliance with the applicable risk-based performance standard. During the inspection, we observed that the facility made this documentation and the appropriate personnel available to answer ISCD inspector questions on the security training the facility held during the prior year. We also observed inspectors verifying that existing measures, such as the facility's fence barrier, were still present and not compromised with breaches. In addition, we observed the inspectors testing key cards to the building that housed the chemicals of interest to ensure the cards prevented unauthorized access. Finally, we observed inspectors requesting a demonstration of how the facility's chemicals of interest are delivered to the facility and what controls were in place to monitor third-party contractors during delivery of chemicals of interest.

We also discussed the compliance inspection process with representatives from trade associations that represent facilities covered by CFATS and considered high-risk.[35] Representatives from 7 of the 11 trade

[35] We contacted officials representing 15 trade associations with members regulated by CFATS and who participated in the Chemical Sector Coordinating Council to obtain their perspectives on DHS efforts to communicate lessons learned to CFATS facilities on methods to reduce risk and the compliance inspection process. For the 11 trade associations that responded, we conducted semistructured interviews and the information we obtained from them is not generalizable to the universe of chemical facilities covered by CFATS. However, the information we obtained from them provides insights into DHS's efforts to perform outreach and seek feedback on the CFATS program.

associations that we spoke with stated that ISCD's implemented changes have improved the compliance inspection process since the program's inception. Specifically, representatives from three trade associations stated that ISCD inspectors' efforts to increase communication with facilities, including preinspection phone calls and compliance assistance visits, have increased the ability of facilities to ensure they are compliant with their approved site security plan. However, representatives from 3 of the 11 trade associations we spoke with also noted some issues with the compliance inspection process. Specifically, officials from these 3 associations stated that ISCD inspectors inconsistently apply the risk-based performance standards relative to the measures the facilities implemented. Some of this inconsistency may be due, in part, to the flexibility inherent in the risk-based performance standards which give facilities the discretion or latitude to tailor security based on conditions and circumstances. For example, the amount and type of chemicals of interest may vary by facility, so some facilities may require additional security measures be put in place to ensure protection of these chemicals. In addition, facilities vary by geographic location, which may affect the measures the facility needs to implement to protect the chemicals of interest from potential theft or diversion.

DHS officials stated that they believe any perceived inconsistency is due to the flexibility in application of the risk-based performance standards and the variety of facility conditions that contribute to the appropriateness of different security measures. Officials explained that, for example, inspectors would likely recommend that a large campus-type facility not invest in a perimeter fence line but instead utilize asset-based barriers to satisfy the performance standards. Officials noted that facilities can choose to employ security measures which best fit their specific situation and can request that inspectors provide multiple options for their consideration.

DHS's Methodology for Measuring Changes in Facility Site Security Does Not Reflect Reduction in Vulnerability

ISCD developed its performance measure methodology for the CFATS program in order to evaluate security changes made by high-risk chemical facilities, but the methodology does not measure the program's impact on reducing a facility's vulnerability to an attack. In 2015 we found that while ISCD's performance measure for the CFATS program was intended to reflect security measures implemented by facilities and the overall impact of the CFATS regulation on facility security, the metric did not solely capture security measures that are implemented by facilities and verified by ISCD.[36] We recommended that DHS develop a performance measure that includes only planned security measures that have been implemented and verified. In response to our finding and recommendation, ISCD's performance measure requires that ISCD officials verify that planned measures have been implemented in accordance with the approved site security plan (or alternative security program) by compliance inspection or other means before inclusion in the performance measure calculation.

ISCD has since decided to develop a new methodology and performance measure for the CFATS program. In 2016, ISCD began development of an approach called the guidepost-based site security plan scoring methodology. ISCD officials stated they plan to use the methodology to evaluate the security measures a facility implemented from initial state— when a facility submits its initial site security plan—to the facility's approved security plan, according to ISCD officials. Officials stated that once implemented, the methodology's resulting performance measure will be maintained internally and, if approved, may be used to satisfy the program's reporting requirements consistent with the Government Performance and Results Act (GPRA) and included in DHS's Annual Performance Report.[37]

---

[36] GAO-15-614.

[37] See generally Government Performance and Results Act of 1993, Pub. L. No. 103-62, 107 Stat. 285 (1993) (GPRA) and GPRA Modernization Act of 2010, Pub. L. No. 111-352, 124 Stat. 3866 (2011) (updating GPRA).

The methodology organizes a facility's security measures based on five guideposts.[38] Using the five guideposts as a framework, the security measures a facility reports in its site security plan are evaluated by ISCD under the applicable guidepost to determine the level of security performance. For example, the plan contains a question on whether a facility has a perimeter fence barrier and if so, what type, such as a chain link fence, metal fence, or vinyl fence. ISCD uses the facility's responses to assign a numerical value that indicates the level of security performance for the type of fence a facility uses as a perimeter barrier. The scores of the five guideposts are then aggregated and the resulting score represents the site security plan score for a facility. Officials stated that a facility's site security plan score is developed when the facility submits its initial site security plan and again when ISCD approves its site security plan and the facility has completed the CFATS inspection process.

ISCD officials stated the purpose of the methodology is to measure the increase in security attributed to the CFATS program and stated that the methodology is not intended to measure risk reduction. As a result, the methodology and resulting performance metric do not reflect the program's impact on reducing a facility's vulnerability to an attack. While ISCD officials stated the program is exploring how to use the site security plan scores of a facility, this methodological approach may provide ISCD an opportunity to begin assessing how vulnerability is reduced and, by extension, risk is lowered, not only for individual facilities but for the program as a whole. The NIPP calls for evaluating the effectiveness of risk management efforts by collecting performance data to assess progress in achieving identified outputs and outcomes. The purpose of the CFATS program is to ensure facilities have security measures in place to reduce the risks associated with certain hazardous chemicals and to prevent these chemicals from being exploited in a terrorist attack. A measure that reflects risk reduction may include how the CFATS inspection process measures the reduction of one element of risk— vulnerability—of high-risk facilities to a terrorist attack. ISCD officials stated that challenges exist with

[38] The five guideposts are (1) detection; (2) delay; (3) response and mitigation; (4) cyber security; and (5) security management.

incorporating vulnerability into the measure's methodology, such as how to accurately measure a facility's vulnerability to an attack before the facility started the CFATS inspection process.

We recognize challenges ISCD might face in incorporating vulnerability into its scoring methodology. In our prior work, we acknowledged that assessing the benefits of a program—such as reducing a high-risk facility's vulnerability to an attack—is inherently challenging because it is often difficult to isolate the impact of an individual program on behavior that may be affected by multiple other factors.[39] However, ISCD could take steps to evaluate vulnerability reduction resulting from the CFATS compliance inspection process. For example, because facilities conduct their own vulnerability assessments when developing their site security plan for submission to ISCD, ISCD could establish a vulnerability baseline score when it evaluates a facility's security measures during its initial review of the facility's plan. ISCD could then use this baseline score as the starting point for assessing any reduction in vulnerability that ISCD can document that has occurred as a result of security measures implemented by the facility during the compliance inspection process.[40] As the CFATS program continues to mature and ISCD begins its efforts to assign scores to facility site security plans, incorporating assessments of reductions in vulnerability at individual facilities and across the spectrum of CFATS facilities as a whole would enable ISCD to better measure the impact of the CFATS compliance inspection process on reducing risk and increasing security nationwide.

---

[39] See, for example GAO, Aviation Security: A National Strategy and Other Actions Would Strengthen TSA's Efforts to Secure Commercial Airport Perimeters and Access Controls, GAO-09-399 (Washington, D.C.: Sept. 30, 2009).

[40] As part of its determination of a facility's risk tier, ISCD requires that a facility complete and submit a Security Vulnerability Assessment. ISCD reviews the security vulnerability assessment provided by the facility to confirm and notify the facility as to whether the facility remains categorized as high-risk and, if so, about its final placement in one of the four tiers.

First Responders and Emergency Planners May Not Have Information Needed to Respond to Incidents at High-Risk Chemical Facilities

We found over 200 chemicals covered by CFATS that may not be included in the chemical inventory information that officials told us they rely on to prepare for and respond to incidents at chemical facilities. ISCD shares some CFATS information with state and local officials, including access to CFATS facility-specific data via a secure portal; however, this portal is not widely used at the local level by first responders and emergency planners.

First Responders and Emergency Planners May Not Have Sufficient Information to Prepare for and Respond to Incidents at High-Risk Chemical Facilities

First responders and emergency planners may not have the necessary information to prepare for and respond to incidents at high-risk chemical facilities regulated by the CFATS program. As mentioned earlier, on April 17, 2013, about 30 tons of ammonium nitrate fertilizer—containing a CFATS chemical of interest—detonated during a fire at a fertilizer storage and distribution facility in West, Texas killing 15 people, including 12 first responders, and injuring more than 260 others. This event, among others, prompted the President to issue Executive Order 13650 to improve chemical facility safety and security in coordination with owners and operators.[41] The Executive Order established a Chemical Facility Safety

[41] See Exec. Order No. 13650 (Aug. 1, 2013), 78 Fed. Reg. 48,028 (Aug. 7, 2013). See also Hearing on Oversight of Federal Risk Management and Emergency Programs to Prevent and Address Chemical Threats, Including the Events Leading up to the Explosions in West, Texas and Geismar, Louisiana, Before the Senate Comm. on Environment and Public Works, 113th Cong., 1st Sess., June 27, 2013 (statement of Rafael Moure-Eraso, Chairman, Chemical Safety and Hazard Investigation Board (Chemical Safety Board). The Chemical Safety Board is an independent federal safety board charged with investigating chemical accidents.

and Security Working Group and included directives for the working group to, among other things, improve operational coordination with state, local, and tribal partners. The working group created a federal plan of action consisting of actions to improve the safety and security of chemical facilities.

One key element of this plan focused on the Emergency Planning and Community Right-to-Know Act of 1986 (EPCRA), which was intended to encourage and support emergency planning efforts at the state and local levels.[42] In accordance with EPCRA, state and local entities, such as Local Emergency Planning Committees (LEPCs)—consisting of representatives including local officials and planners, facility owners and operators, first responders, and health and hospital personnel, among others—were created.[43] These LEPCs were designed to (1) prepare for and mitigate the effects of a chemical incident and (2) ensure that information on chemical risks in the community is provided to first responders and the public. The working group acknowledged there was a need to share data with representatives of these state and local entities to enable them to identify gaps and inconsistencies in their existing information that could reveal previously unknown risks in their communities. For facilities subject to EPCRA requirements, this data is to include, among other things, information about chemicals stored or used at the facility for which facilities are required to submit an emergency and hazardous chemical inventory form to these state and local entities.[44] The working group's federal plan also included a DHS commitment to share certain CFATS

---

[42] See Pub. L. No. 99-499, tit. III, 100 Stat. 1613, 1728 (1986); 42 U.S.C. §§ 11001-11050. See also 40 C.F.R. pts. 355, 370.

[43] See 42 U.S.C. § 11001. In addition to LEPCs, State Emergency Response Commissions (SERCs), Tribal Emergency Response Commissions (TERCs), and Tribal Emergency Planning Committees (TEPCs) were established in accordance with EPCRA. According to the Chemical Facility Safety and Security Working Group, strengthening SERC, TERC, LEPC, and TEPC capabilities is critical to improving chemical facility safety and security.

[44] Under Section 312 of the Emergency Planning and Community Right-to-Know Act of 1986 (EPCRA), facilities are required to submit an emergency and hazardous chemical inventory form—referred to as a Tier II form. See 42 U.S.C. § 11022. The purpose of this form is to provide state and local officials and the public with specific information on potential hazards. This includes the locations and amount of hazardous chemicals present at a facility during the previous calendar year.

data elements with first responders, state agencies and LEPCs to help communities identify and prioritize risks and develop a contingency plan to address those risks while acknowledging that access to certain sensitive portions of CFATS data will remain restricted to officials with a "need-to-know" so as to appropriately balance security risks.

In our interviews with 15 LEPCs—whose jurisdictions include 373 high-risk chemical facilities regulated by the CFATS program—we found that officials rely on information reported on EPCRA chemical inventory forms to prepare for and respond to incidents at CFATS facilities.[45] These officials may not have sufficient information to respond to emergencies at CFATS facilities because EPCRA reporting requirements may not cover some of the chemicals covered under the CFATS program.[46] Specifically, we analyzed the chemicals covered by both CFATS and EPCRA's reporting requirements and found there are over 200 CFATS chemicals of interest that, depending upon state reporting guidelines, may not be covered by EPCRA reporting requirements. Several of these chemicals may require specific response techniques to minimize the risk of injury or death to first responders and the surrounding community. For example, in the event of fire, aluminum powder, a chemical not subject to EPCRA reporting requirements but regulated under CFATS, produces flammable gases when in contact with water and requires responders to instead use a dry chemical or sand to extinguish the fire. Based on our analysis of tiered CFATS facilities, we estimate that about 32 percent of these high-risk facilities possess at least one chemical that may not be covered by EPCRA reporting requirements.[47]

In addition, we found these LEPCs may lack information on the CFATS facilities in their jurisdictions. Specifically, officials representing

---

[45] We interviewed officials representing 15 LEPCs out of more than 3,000 known LEPCs. We selected LEPCs from different states to include counties from among those with the highest number of facilities in each state. The number of high-risk CFATS facilities located in each LEPC ranges from a low of 11 to a high of 88 across our sample.

[46] Some states have lower reporting thresholds for chemicals subject to EPCRA reporting requirements or expanded lists of chemicals that must be reported.

[47] We selected a generalizable random sample of all 3,539 tiered CFATS facilities as of January 8, 2018. Based on this sample, we estimated that 31.7 percent of the facilities possessed at least one chemical that may not be covered by EPCRA; this estimate has a margin of error at the 95 percent confidence level of plus or minus 5 percentage points.

11 of the 15 LEPCS we interviewed said they were not aware of which facilities in their jurisdiction were regulated by the CFATS program. Of these 11 LEPCs, officials from 8 LEPCs stated it would be very helpful or critical to know this information and officials from 2 LEPCs stated it would be somewhat helpful.[48] According to these officials, this information would assist LEPCs, some of which have hundreds of facilities in their jurisdiction, to prioritize the most significant facilities for additional planning or scheduling of drills and exercises. Additionally, officials representing 5 LEPCs stated they were not aware of the differences between CFATS chemicals of interest and those chemicals subject to EPCRA reporting requirements. These LEPC officials stated that, among other things, it is critical to have a comprehensive understanding of all chemicals at a facility and that this information is very important for emergency responders to be aware of when responding to an incident.

ISCD Could Take Additional Action to Share Information about High-Risk Facilities with First Responders and Emergency Planners

Consistent with the CFATS Act of 2014, ISCD is to play a role in ensuring that first responders and emergency planners are properly prepared for and provided with the situational awareness needed to respond to security incidents at high-risk chemical facilities.[49] While the CFATS Act of 2014 does not specifically require that information be shared directly with first responders, ISCD has taken steps to share CFATS information with state and local officials to help ensure that first responders are prepared to respond to such security incidents.[50] These steps

[48] Officials from 1 LEPC stated they did not know how helpful this information would be.

[49] See 6 U.S.C. § 623.

[50] Specifically, the statute provides that the Secretary shall provide to state, local, and regional fusion centers and state and local government officials, as the Secretary determines appropriate, such information as is necessary to help ensure that first responders are properly prepared and provided with the situational awareness needed to respond to security incidents at covered chemical facilities. See 6 U.S.C. § 623(c)(1); see also 6 U.S.C. 124h(j)(1) (defining "fusion center"). The statute further provides that the Secretary shall

include, among other things, ensuring that facilities are developing and exercising an emergency plan to respond to security incidents internally and with assistance of local law enforcement and first responders. Planning and training are important to ensure that facility personnel, onsite security, law enforcement, and first responders are ready to respond to external and internal security incidents. Additionally, these planning activities and relationships with first responders can assist in reducing the impact of these incidents. According to ISCD officials, to verify compliance with this requirement, ISCD inspectors validate facility outreach to first responders, such as local law enforcement and fire departments, through review of facility documentation, including emails with first responders, records of drills, and logs of meetings and tours, or through direct contact with the local first responders by the inspection team.

In addition, the Executive Order 13650 working group sought to, among other things, strengthen community planning and preparedness and ensure that first responders and emergency planners are aware of the risks associated with hazardous chemicals in their communities. Included was a goal to increase information-sharing with communities that are near chemical facilities. In a May 2014 report, this working group identified certain information, including the name and quantity of chemicals at a facility, as the most helpful to first responders and emergency planners.[51] This information is intended to enable emergency planners to conduct an analysis to identify gaps and inconsistencies in their existing information that could reveal previously unknown risks in their communities.

ISCD has taken action to ensure first responders and emergency planners have access to CFATS data. For example, in response to Executive Order 13650, ISCD shares CFATS data through the

---

disseminate such information through a medium or system determined by the Secretary to be appropriate to ensure the secure and expeditious dissemination of such information to necessary selected individuals. See § 623(c)(2).

[51] Actions to Improve Chemical Facility Safety and Security—A Shared Commitment, Report for the President (May 2014). See Exec. Order No. 13650, 78 Fed. Reg. at 48,029, § 2(c) (directing the submission of a status report within 270 days of the date of the Executive Order).

Infrastructure Protection (IP) Gateway.[52] This online portal contains critical infrastructure data and analytic tools, including data on covered CFATS facilities, for use by federal officials, state, local, tribal, and territorial officials, and emergency response personnel.[53] CFATS data available in the IP Gateway includes, among other things, facility name, location, risk tier, and chemicals on-site and is accessible to authorized federal and other state, local, tribal, and territorial officials and responders with an established need-to-know.[54] The IP Gateway provides these officials and responders access to CFATS facility-specific information that may be unreported on EPCRA chemical inventory forms. This CFATS facility-specific information can help ensure these groups are properly prepared to respond to incidents at high-risk chemical facilities in their jurisdictions.

While the IP Gateway is a mechanism for sharing names and quantities of chemicals at CFATS high-risk facilities with first responders and emergency planners, we found it is not widely used by officials at the local level. ISCD told us that in May 2018 they published three revised fact sheets and included information on the IP Gateway in presentation materials that officials told us was intended to increase promotion and use of the IP Gateway. However, according to DHS, there are 14 accounts categorized at the local level whose access to the IP Gateway layer includes the names and quantities of chemicals at CFATS facilities. A local account indicates the individual with access is a county- or city-level

---

[52] In addition to the IP Gateway, as a result of work by the Executive Order 13650 working group, some CFATS information was added to the Computer-Aided Management of Emergency Operations (CAMEO) database. CAMEO, developed by the Environmental Protection Agency and the National Oceanic and Atmospheric Administration, is a system of software applications used to plan for and respond to chemical emergencies. The CAMEO database is not a complete source of information for first responders to use in an emergency situation because, as a publicly available database, it does not contain specific facility data or other sensitive security information such as which facilities in their jurisdictions hold these chemicals or in what quantities.

[53] The IP Gateway, hosts DHS's Office of Infrastructure Protection's facility database, which records, among other things, IP's assessments and other interactions with facilities. The IP Gateway portal is restricted and allows authorized users to obtain, post, and exchange information and access common resources, particularly critical infrastructure information, including security survey data.

[54] Need-to-know as determined by ISCD.

employee or contractor.[55] Additionally, while not generalizable to all LEPCs, officials representing 7 of the 15 LEPCs we interviewed were not aware of the IP Gateway and officials representing 13 of the 15 LEPCs stated that they do not have access to CFATS information within the IP Gateway. Of the 13 officials that reported they did not have access, 11 said that it would be helpful or critical to have access for several reasons. Specifically, officials representing these LEPCs stated that this information would assist them to better prepare and respond to incidents and help emergency planners prioritize the most critical sites among the thousands of facilities that they oversee.

According to DHS officials, their outreach plan, developed in March 2015, specifically addresses regular engagement with LEPCs, among other groups.[56] However, these officials acknowledged that information may not be reaching some state and local officials due to a number of factors, including the large number of LEPCs and first responders across the country, and changes in the level of LEPC activity and personnel over time. While we recognize these challenges, providing first responders and emergency planners access to CFATS facility-specific information, including the name and quantity of chemicals at a facility, can help ensure these groups are properly prepared to respond to incidents at high-risk chemical facilities in their jurisdictions. The NIPP states that agencies should share actionable and relevant information across the critical infrastructure community—including first responders and emergency planners—to build awareness and enable risk-informed decision making as these stakeholders are crucial consumers of risk information.[57]

---

[55] Account requests for access to the IP Gateway are made via a web-based registration form that asks the individual requesting access to identify the type of employee they are. Options include: Federal, State, Local (City/County), and Tribal/Territory.

[56] DHS officials stated that by the end of fiscal year 2017, DHS had conducted nationwide outreach with more than 1,000 state and local offices and 1,400 LEPCs across the country. They further stated that they plan to conduct outreach with state officials once per year and with LEPCs and TEPCs at least once every three years, as applicable. DHS officials also stated that they have developed relationships with national, local, and first responder organizations to further leverage their networks and outreach activitie.

[57] See DHS, National Infrastructure Protection Plan (Washington, D.C.: June 2006). DHS issued the NIPP in response to the Homeland Security Act of 2002, as amended, and other authorities and directives. See, e.g., Pub. L. No. 107-296, § 201(d)(5), 116 Stat. 2135, 2146 (2002); 6 U.S.C. § 121(d)(5). DHS updated the NIPP in January 2009 to include a greater

Additionally, the 2015 Emergency Services Sector-Specific Plan, an Annex to the 2013 NIPP, further calls for engaging with local emergency planning organizations, such as LEPCs, to enhance information-sharing and analytical capabilities for incident planning, management, and mitigation between stakeholders. The IP Gateway is one way through which ISCD can share CFATS facility-specific information, including the name and quantity of chemicals at high-risk facilities with first responders and emergency planners. As discussed earlier, although ISCD is not required to share CFATS facility-specific information directly with first responders, this information is critical to prepare for and respond to incidents at high-risk chemical facilities and to protect them and their communities from injury or death. By exploring ways to improve information-sharing of CFATS facility-specific data, such as promoting wider use of the IP Gateway among first responders and emergency planners, DHS will have greater assurances that the emergency response community has access to timely information about high-risk chemical facilities.

CONCLUSION

DHS, through ISCD, has made improvements to the CFATS program. ISCD has taken action to strengthen its processes for verifying the accuracy of data it uses to identify high-risk chemical facilities, revised its risk assessment methodology to more accurately identify and assign high-risk chemical facilities to tiers, and has nearly completed its efforts to apply this new methodology to facilities covered by CFATS. Furthermore, ISCD has conducted an increased number of compliance inspections and continues to make changes to improve the efficiency of the inspection process. While ISCD has developed a new methodology and performance

emphasis on resiliency. See DHS, National Infrastructure Protection Plan, Partnering to Enhance Protection and Resiliency (Washington, D.C.: January 2009). DHS further updated the NIPP, which is now called the National Plan, in December 2013. See DHS, NIPP 2013, Partnering for Critical Infrastructure Security and Resilience (Washington, D.C.: December 2013).

measure for the CFATS program in order to evaluate security changes made by high-risk chemical facilities, we found that the methodology and metric do not reflect the program's impact on reducing a facility's vulnerability to an attack. ISCD may have an opportunity to explore how reductions in vulnerability at individual facilities resulting from the CFATS compliance inspection process could be used to develop an overall measure of the performance of the CFATS program in reducing risk and increasing security nationwide. Such a measure would be consistent with the NIPP, which calls for evaluating the effectiveness of risk management efforts by collecting performance data to assess progress in achieving identified outputs and outcomes. Moving forward, ISCD could also take additional actions to ensure information about high-risk chemical facilities is shared with first responders and emergency planners. During our review, we found that local emergency responders may not have the information they need to adequately respond to incidents at CFATS facilities; a situation that could expose them and their communities to potentially life-threatening situations. While the IP Gateway is a mechanism for sharing names and quantities of chemicals at high-risk facilities with first responders and emergency planners, we found it is not widely used by officials at the local level. The NIPP states that agencies should share actionable and relevant information across the critical infrastructure community—including first responders and emergency planners—to build awareness and enable risk-informed decision making, as these stakeholders are crucial consumers of risk information. By improving information-sharing with first responders and emergency planners, such as promoting access to and wider use of the IP Gateway, DHS will have greater assurances that the emergency response community has access to timely information about high-risk chemical facilities that could help protect them from serious injury or death.

Recommendations for Executive Action

We are making the following two recommendations to DHS:

The Director of ISCD should incorporate vulnerability into the CFATS site security scoring methodology to help measure the reduction in the vulnerability of high-risk facilities to a terrorist attack, and use that data in assessing the CFATS program's performance in lowering risk and enhancing national security. (Recommendation 1)

The Assistant Secretary for Infrastructure Protection, in coordination with the Director of ISCD, should take actions to encourage access to and wider use of the IP Gateway and explore other opportunities to improve information-sharing with first responders and emergency planners. (Recommendation 2)

AGENCY COMMENTS AND OUR EVALUATION

We provided a draft of this chapter to DHS for review and comment. DHS provided written comments, which are reproduced in full in appendix I, and technical comments, which we incorporated as appropriate. In its comments, DHS concurred with both recommendations and outlined efforts underway or planned to address them.

Regarding the first recommendation that ISCD should incorporate vulnerability into the CFATS site security scoring methodology to help measure the reduction in the vulnerability of high-risk facilities and use that data to further assess the CFATS program's performance in lowering risk and enhancing national security, DHS concurred but noted that developing a system that could numerically evaluate vulnerabilities will be challenging. DHS stated that implementing the recommendation would likely require, among other things, revising the regulatory language describing CFATS vulnerability assessments and updating tools used to gather them, potentially creating a significant burden on both industry and government. DHS added that its new proposed performance metric, described earlier in this chapter, demonstrates the enhancement to national security resulting from the CFATS program and, by extension, the program's impact on vulnerability and overall risk.

As stated earlier, we recognize challenges ISCD might face in incorporating vulnerability into its scoring methodology. In our prior work, we acknowledged that assessing the benefits of a program—such as reducing a high-risk facility's vulnerability to an attack—is inherently challenging because it is often difficult to isolate the impact of an individual program on behavior that may be affected by multiple other factors. However, in order to fully implement this recommendation, ISCD needs to consider steps it can take to evaluate vulnerability reduction resulting from the CFATS compliance inspection process without revisions to the regulation or by creating a significant burden on both industry and government. We noted, for example, that ISCD could establish a vulnerability baseline score when it evaluates a facility's security measures during its initial review of the facility's site security plan. ISCD could then use this baseline score as the starting point for assessing any reduction in vulnerability that ISCD can document that has occurred as a result of security measures implemented by the facility during the compliance inspection process. As the CFATS program continues to mature and ISCD begins its efforts to assign scores to facility site security plans, incorporating assessments of reductions in vulnerability at individual facilities and across the spectrum of CFATS facilities as a whole would enable ISCD to better measure the impact of the CFATS compliance inspection process on reducing risk and increasing security nationwide.

Regarding the second recommendation that the Office of Infrastructure Protection and ISCD take actions to encourage access to and wider use of the IP Gateway and explore other opportunities to improve information-sharing with first responders and emergency planners, DHS stated that it has various outreach activities underway, among other information-sharing efforts, to either directly share or ensure that high-risk chemical facilities are sharing CFATS information with first responders and emergency planners. DHS added that, to continue these efforts and to encourage better utilization of the IP Gateway, it will ensure contact is made with LEPCs representing the top 25 percent of CFATS high-risk chemical facilities no later than the end of the second quarter of fiscal year 2019. While the outreach and information-sharing efforts DHS described are a step in the right direction, in order to fully implement this recommendation it is critical that the intent of any actions taken is to ensure that all first responders and emergency planners with a need-to-know are provided with timely access to CFATS facility-specific information in their jurisdictions. This information should include the name and quantity of chemicals at a facility so as to help these groups be properly prepared to respond to incidents at high-risk chemical facilities and to minimize the risk of injury or death to first responders and the surrounding community. Furthermore, it is important that these actions are focused on ensuring that this CFATS facility-specific information is shared with first responders and emergency planners representing the entirety of CFATS facilities determined to be high-risk, not just those that represent the top 25 percent of CFATS high-risk facilities.

Chris P. Currie Director
Homeland Security and Justice

APPENDIX I: COMMENTS FROM THE DEPARTMENT OF HOMELAND SECURITY

U.S. Department of Homeland Security
Washington, DC 20528

Homeland
Security

July 27, 2018

Chris P. Currie
Director, Homeland Security and Justice
U.S. Government Accountability Office
441 G Street, NW
Washington, DC 20548

Re: Management's Response to Draft Report GAO-18-538, "CRITICAL INFRASTRUCTURE PROTECTION: DHS Should Take Actions to Measure Reduction in Chemical Facility Vulnerability and Share Information with First Responders"

Dear Mr. Currie:

Thank you for the opportunity to review and comment on this draft report. The U.S. Department of Homeland Security (DHS) appreciates the U.S. Government Accountability Office's (GAO) work in planning and conducting its review and issuing this report.

The Department is pleased to note GAO's positive recognition of the substantial progress DHS has made in developing and implementing the revised risk tiering methodology used for the Chemical Facility Anti-Terrorism Standards (CFATS) regulation as well as the progress made in conducting CFATS compliance inspections. For example, the report acknowledges progress made (1) completing security plan approvals and compliance inspections, effectively eliminating the backlog of security plan approvals in 2016, approximately six years ahead of previous GAO estimates, and (2) significantly increasing the number of compliance inspections, with more than 3,500 compliance inspections completed as of May 2018. DHS remains committed to ensuring not only that high-risk chemical facilities are identified and implementing appropriate security measures, but also that the CFATS program is being administered in a fair, efficient, and effective manner.

The draft report contained two recommendations with which the Department concurs. Attached find our detailed response to the recommendations. Technical comments were previously provided under separate cover.

Again, thank you for the opportunity to review and comment on this draft report. Please feel free to contact me if you have any questions. We look forward to working with you in the future.

Sincerely,

JIM H. CRUMPACKER, CIA, CFE
Director
Departmental GAO-OIG Liaison Office

Attachment: Management Response to Recommendations Contained in GAO 18-538

GAO recommended that:

Recommendation 1: The Director of National Protection & Programs Directorate's (NPPD) Office of Infrastructure Protection, specifically the Infrastructure Security Compliance Division (ISCD) incorporate vulnerability into the Chemical Facility Anti-Terrorism Standards Compliance (CFATS) site security scoring methodology to help measure the reduction in the vulnerability of high-risk facilities to a terrorist attack, and use that data in assessing the CFATS program's performance in lowering risk and enhancing national security.

Response: Concur. NPPD's Office of Infrastructure Protection (IP) previously identified that the Department's current CFATS performance metric, reported in accordance with the Government Performance and Results Act, does not adequately demonstrate the security impact of the CFATS program. To address this deficiency, NPPD/IP has begun drafting and testing a proposed metric that program officials and leadership believes better reflects the program's effectiveness in enhancing national security and reducing national risk. The new metric is intended to (1) evaluate the progress individual facilities have made in enhancing their security while part of the CFATS program, and (2) will be based on a comparison of the security measures employed by CFATS-covered chemical facilities upon first entering the program with the improved and enhanced security measures contained in their approved CFATS security plans. Moreover, the information obtained by this new metric will then be aggregated to determine the average percent increase in the security score of facilities' security plans, resulting in a representation of the increase in the security posture across the facility population.

It is important to point out, however, that implementing this specific recommendation will be challenging. This is largely due to the fact that the information currently collected by the CFATS tools on facility vulnerabilities is gathered from facilities, and does not include an assessment conducted by the Department. Therefore, the information is not in a format that would support such a numerical calculation. More specifically, much of the vulnerability information submitted by facilities through the current suite of CFATS tools is written text which the Department uses to identify potential security gaps that need to be addressed by the security plan rather than provision of a vulnerability score. To develop a system that could numerically evaluate vulnerabilities likely would require revising the regulatory language describing CFATS vulnerability assessments, modifying CFATS process, and updating tools used to gather vulnerability assessments. This would be a significant burden on both industry and government and NPPD does not believe this would result in a better measure for evaluating the security enhancing effectiveness of the CFATS program, compared to the new performance measure the Department intends to implement.

NPPD/IP believes this new metric will meet the spirit and intent of GAO's recommendation by directly demonstrating the enhancement to national security resulting from the CFATS program, and, by extension, demonstrating the program's impact on vulnerability and overall risk. The Department intends to begin reporting this measure shortly after the end of the first quarter of FY 2019.

Estimated Completion Date (ECD): January 31, 2019.

Recommendation 2: The Assistant Secretary for Infrastructure Protection, in coordination with the Director of ISCD, take action to encourage access to and wider use of the IP Gateway and explore other opportunities to improve information-sharing with first responders and emergency planners.

Response: Concur. DHS agrees that first responders and emergency planners should be aware of the potentially dangerous chemicals located in their community and the facilities that possess them. As such, IP has undertaken numerous efforts to share CFATS information with emergency planners and first responders with a "need-to-know." For example, IP regularly engages with the Emergency Services Sector Government and Sector Coordinating Councils to learn about opportunities for coordination and to provide updates on the CFATS program. DHS also engages with State, local, tribal, and territorial agencies and departments, including first responders, State Emergency Response Commissions, Tribal Emergency Response Commissions, Local Emergency Planning Committees (LEPCs), and Tribal Emergency Planning Committees (TEPCs). Moreover, during FY 2017, DHS conducted nationwide outreach to more than 1,200 State and local offices and over 1,400 LEPCs/TEPCs across the 50 U.S. states, District of Columbia, and 9 U.S. territories, and, as described in the CFATS Outreach Implementation Plan, and intends to continue to conduct outreach with State officials once a year and with LEPCs/TEPCs at least once every three years, as applicable.

The Department also makes CFATS data available to first responders via the Infrastructure Protection (IP) Gateway Executive Order 13650 Portal. First responders and Federal, State, local, tribal, and territorial, governments who have a "need-to-know" are now able to view facilities in their area of responsibility, which improves data-sharing and emergency planning efforts. Instances where first responders have been reluctant to request an IP Gateway account or prefer access to the information via other means, NPPD/IP has provided electronic or hard copy lists of facilities and their chemical information to authorized individuals with a need to know. Additionally, CFATS information was added to the Computer-Aided Management of Emergency Operations Chemicals database, the National Oceanic and Atmospheric Administration's online tool which emergency responders and planners can use to get response recommendations and predict hazards such as explosions or toxic fumes. Recently, IP also revised three fact sheets and an outreach presentation to include information on the IP Gateway and how to request access to it.

In addition to the outreach activities described above, NPPD/IP, through the CFATS Risk-Based Performance Standard 9–Response, ensures that high-risk chemical facilities have regular and recurring contact with their local first responders to ensure they are properly prepared to respond to incidents at their facilities. IP believes that this effort is the most effective way to get information to first responders as it involves direct communication between the high-risk chemical facilities and their local responders. While DHS cannot require first responders access the IP Gateway or respond to facility requests for visits, DHS does believe that we are effectively implementing information sharing as part of the CFATS program. To continue these efforts and to encourage better utilization of the IP Gateway, DHS will ensure contact is made with LEPCs representing the top 25 percent of CFATS high-risk chemical facilities no later than the end of the second quarter of FY 2019. ECD: March 31, 2019.

In: Top 5 Reports and Testimonies … ISBN: 978-1-53615-511-2
Editors: C. Johnston and W. Payne

Chapter 5

FEDERAL EMPLOYEE MISCONDUCT: ACTIONS NEEDED TO ENSURE AGENCIES HAVE TOOLS TO EFFECTIVELY ADDRESS MISCONDUCT*

United States Government Accountability Office

ABBREVIATIONS

ADR	alternative dispute resolution
CBA	collective bargaining agreement
CFO Act	Chief Financial Officers Act
CHCO	Chief Human Capital Officer
CSRA	Civil Service Reform Act of 1978
EEOC	Equal Employment Opportunity Commission
EHRI	Enterprise Human Resources Integration database

* This is an edited, reformatted and augmented version of United States Government Accountability Office; Report to Congressional Requesters, Publication No. GAO-18-48, dated July 2018.

EPA	Environmental Protection Agency
MSPB	Merit Systems Protection Board
NOA	nature of action
OI	Office of Inspector General Office of Investigations
OPM	Office of Personnel Management
OSC	Office of Special Counsel
PFR	petition for review
SF-50	Standard Form 50

WHY GAO DID THIS STUDY

Misconduct is generally considered an action by an employee that impedes the efficiency of the agency's service or mission. Misconduct incidents can affect other aspects of employee morale and performance and impede an agency's efforts to achieve its mission.

GAO was asked to examine how executive branch agencies address employee misconduct. This chapter (1) describes the process agencies are required to follow in responding to employee misconduct; (2) identifies alternative approaches to the formal process that agencies can use and assesses what factors affect agencies' responses to misconduct; (3) describes trends in removals and other adverse actions resulting from misconduct; and (4) identifies key practices agencies can use to help them better prevent and address misconduct. To address these objectives, GAO reviewed relevant sections of title 5 of the U.S.C; analyzed MSPB and OPM data, and interviewed, among others, agency officials and subject-matter experts.

WHAT GAO RECOMMENDS

GAO recommends that OPM, working with the Chief Human Capital Officers Council, (1) take steps to improve the quality of data collected on misconduct; (2) leverage lessons learned to help agencies address misconduct; and (3) improve guidance on training supervisors and human resources staff on addressing misconduct. OPM partially concurred with two recommendations, and disagreed with the first, stating that its guidance has been successfully relied upon by agencies. GAO maintains the action is needed to help strengthen oversight.

WHAT GAO FOUND

Chapter 75 of title 5 of the U.S. Code specifies the formal legal process that most agencies must follow when taking adverse actions, i.e., suspensions, demotions, reductions in pay or grade, and removals, for acts of employee misconduct. Chapter 75 details the built-in procedural rights certain federal employees are entitled to when faced with adverse actions.

Depending on the nature of misconduct, an agency may use utilize alternative discipline approaches traditionally used in government to correct behavior. Alternative discipline is an approach to address misconduct that is available to agencies in lieu of traditional penalties (e.g., letters of reprimand and suspensions of 14 days or less). An example is a last chance agreement, whereby an employee recognizes the agency's right to terminate him or her should another act of misconduct occur.

Based on the data collected by the Office of Personnel Management (OPM), agencies formally discipline an estimated 17,000 employees annually under Chapter 75, or less than 1 percent of the federal workforce, for misconduct. Based on OPM data, in 2016, agencies made 10,249 suspensions, 7,411 removals, and 114 demotions for misconduct. However, because of weaknesses in OPM's data on employee misconduct, which is provided by the agencies, OPM is unable to accurately target supervisory training to address misconduct, and decision-makers do not know the full extent or nature of this misconduct.

Key Practices That Can Help Agencies Better Prevent and Respond to Misconduct

Senior Agency Officials Must Set Positive Conduct Examples (tone at the top)	Senior leaders must exhibit positive workplace behavior as an example to agency employees.
Additional Training Could Help Supervisors Identify and Deal with Misconduct	Conduct on-going training for supervisors and hold them accountable for addressing misconduct in a timely manner when it occurs
Internal Collaboration is Key to Effectively Address Employee Misconduct	Maintaining effective lines of communication and collaboration with the human resources office staff, line-level management, and agencies' legal counsel
Set Clear Expectations and Engaging Employees	Setting and communicating clear rules and expectations regarding employee conduct and assuring that employees conform to any applicable standards of conduct

Source: GAO analysis of expert testimony. | GAO-18-48.

Key lessons learned can help agencies better prevent and respond to misconduct. For example, tables of penalties provide a list of the infractions committed most frequently by agency employees, along with a suggested range of penalties for each to ensure consistent treatment for similar offenses. However, not all agencies have a table of penalties, including OPM, nor are agencies required by statute, case law or OPM regulations. Subject-matter experts we contacted identified additional promising practices that agencies can use to respond employee misconduct. Some of these are presented above.

Agencies are accountable for providing required training to their managers. However, agency officials and subject-matter experts we interviewed said federal managers may not address misconduct because they are unfamiliar with the disciplinary process, have inadequate training, or receive insufficient support from their human resources offices.

July 16, 2018

The Honorable James Lankford
Chairman

Subcommittee on Regulatory Affairs and Federal Management
Committee on Homeland Security and Governmental Affairs
United State Senate

The Honorable Elijah E. Cummings
Ranking Member
Committee on Oversight and Government Reform
House of Representatives

An average of less than 1 percent of the federal government's 2.1 million employees are formally disciplined for misconduct annually, according to our analysis of the Office of Personnel Management's (OPM) data.[1] While this number is relatively small, even a few cases of employee misconduct can have significant impacts on workplace morale and impede an agency's efforts to achieve its mission. Indeed, according to a 2008 study by the Merit System Protection Board (MSPB), "Misconduct affects not only the employee and the supervisor; it also impacts those who interact with the employee."[2] To that end, it is important for agencies to timely and effectively address cases of employee misconduct while simultaneously respecting employees' procedural and due process rights.

The term "employee misconduct" does not have a general definition in a statute or government-wide regulations. Agencies may elaborate on types of misconduct in handbooks and other internal guidance. However, according to OPM, there is a large body of decisional law by MSPB addressing discipline for employee misconduct in the federal government that contains definitions of various forms of misconduct, such as "insubordination," "excessive absence," and "misuse of government property," as well as instances where specific types of misconduct are referenced concerning appointment into the competitive service.[3]

[1] Formal discipline refers to the legal process under Chapter 75 of Title 5 of the U.S. Code.

[2] Merit Systems Protection Board, *Alternative Discipline: Creative Solutions for Agencies to Effectively Address Employee Misconduct*. (Washington, D.C.: July 2008).

[3] 5 U.S.C. §§ 7301-7363. There are instances in law and regulation where specific types of misconduct are referenced concerning appointment into the competitive service. 5 U.S.C.

According to OPM officials, the lack of a general definition of misconduct gives agencies flexibility in the manner in which they address misconduct. Ultimately, when an agency needs to take an adverse action[4] against an employee for inappropriate workplace behavior, it must do so "for such cause as will promote the efficiency of the service" as provided for in 5 U.S.C. Chapter 75.[5] Such actions could include, but are not limited to, instances of misconduct. The efficiency of the service standard provides managers with maximum flexibility to pursue adverse actions whether the impetus was a conduct issue, a failure to perform, or something else.

Activities that may be considered misconduct can vary from agency to agency. According to agency officials and subject-matter experts we interviewed, examples of misconduct include

- time and attendance infractions;
- intoxication;
- workplace violence;
- physical aggression toward an employee;
- improper use of a government-issued credit card;
- misuse of government equipment (such as viewing pornography or gambling);
- use of public position for private gain; and
- behavior that affects national security.

It should be noted that the law differentiates misconduct from substandard performance.[6] Poor performance can generally be described as

---

Chapter 73 (Suitability, Security, and Conduct) addresses certain types of misconduct of executive branch employees.

[4] Adverse actions can be taken for any reason to promote the efficiency of the service. Such reasons can include, but are not limited to insubordination, excessive absence, and misuse of government property.

[5] 5 U.S.C. § 7503(a); 5 U.S.C. § 7513(a). The purpose of Chapter 75 was to establish procedures that must be followed when an agency determines to pursue certain forms of actions against an employee that would, if taken, have an impact upon their pay, i.e., "adverse actions."

[6] *See* 5 U.S.C. §§ 7501-7514; 5 U.S.C. §§ 4301-4306. According to OPM, poor performance and misconduct often overlap; nevertheless, it is important to recognize the difference between the two. OPM describes misconduct as a failure to follow a workplace rule (e.g., tardiness and absenteeism) and poor performance as the failure of an employee to do his or her job at

an employee's inability to carry out work responsibilities, while misconduct can be described as an employee, willfully or otherwise, violating stated policies or norms. In certain cases, however, employee performance and misconduct can overlap, combining the two issues. Sleeping on the job, for example, is both a failure to abide by norms of conduct an agency would expect of an employee on duty and a failure to perform.

Our prior work on poor performers found that adverse actions, including suspensions, demotions, and removals, take time to resolve and because of a lack of internal support, concerns over litigation and other factors, supervisors may be hesitant to initiate required procedures outlined in the United States Code (hereinafter U.S.C.).[7] Supervisors can address misconduct through alternative discipline approaches which can shorten the timeline and eliminate costly litigation if the employee decides to appeal an adverse action. When used under the right circumstances, alternative discipline may be more efficient and effective than traditional discipline.[8]

You asked us to examine the process for addressing misconduct and to identify any challenges to removing employees for misconduct. Our objectives were to (1) describe the process that agencies are required to follow in responding to employee misconduct in the federal service; (2) identify alternative approaches to the formal legal process that agencies can use to respond to misconduct, and assess what factors affect agencies' responses; (3) describe trends in removals and other adverse actions resulting from misconduct; and (4) identify key practices agencies can use to help them better prevent and address misconduct.

---

an acceptable level. OPM, *Addressing and Resolving Poor Performance: A Guide for Supervisors* (Washington, D.C.: March 2017) https://www.opm.gov/policy-data-oversight/employee-relations/employee-rightsappeals/performance-based-actions/toolkit.pdf.

[7] GAO, *Federal Workforce: Improved Supervision and Better Use of Probationary Periods Are Needed to Address Substandard Employee Performance*, GAO-15-191 (Washington, D.C.: Feb. 6, 2015).

[8] Merit Systems Protection Board, *Alternative Discipline: Creative Solutions for Agencies to Effectively Address Employee Misconduct* (Washington, D.C.: July 2008).

To address the first objective, we reviewed relevant sections of Title 5 chapter 75 of the U.S.C. (hereinafter chapter 75) which contains the statutory process for taking such adverse actions as are necessary to promote the efficiency of the service, which would include adverse actions to address misconduct.[9] We also reviewed OPM regulations implementing chapter 75 adverse action procedures and describing employee grievance and appeal rights.[10] Additionally, we reviewed 5 U.S.C. §§ 7701 and 7702 (hereinafter chapter 77), which contain the statutory process for adverse action appeals to the MSPB and subsequent appeals to the Equal Employment Opportunity Commission (EEOC) when the employee has alleged that an adverse action within the jurisdiction of the MSPB was taken on account of discrimination within the jurisdiction of the EEOC.[11] To provide context for our review of Chapters 75 and 77, we interviewed current and former practitioners, subject-matter experts, and academics. We selected our subject-matter experts based on our guidance for selecting experts, including their depth of experience on this issue and their present and past employment history, the subject-matter experts' practical experience in applying and practicing administrative law related to employee misconduct, and academicians who have conducted research on employee misconduct.

To address the second objective, we interviewed OPM and MSPB officials and reviewed MSPB documents to obtain alternative discipline approaches agencies may use to address employee misconduct. We reviewed OPM regulations and documents to determine the legal framework within which most agencies address employee misconduct in the federal service, including formal procedural and employee appeal rights. To describe and assess factors that affect an agency's response to employee misconduct, we interviewed a panel of chief human capital officers (CHCO) from selected agencies as well as current and former human capital practitioners and other individuals and organizations with

---

[9] *See generally*, 5 U.S.C. §§ 7501-7514. Other adverse that may be taken under Chapter 75 include, for example, a furlough of 30 days or less for the purpose of keeping the agency's spending within appropriation limits.

[10] OPM regulations implementing Chapter 75 are found in 5 C.F.R. subpart D of part 752.

[11] 5 U.S.C. § 7701(a)-(b); 5 U.S.C. § 7702(b).

expertise in working on employee misconduct issues.[12] These interviews helped us to gain insight into the agency perspective on addressing employee misconduct.

To address the third objective, we analyzed data from OPM's Enterprise Human Resources Integration (EHRI) database and MSPB data from fiscal years 2006 to 2016 to describe trends in misconduct-related personnel actions and determine the number of adverse actions that agencies take against employees government-wide.[13] Some agencies are excluded from the EHRI database, such as those in the intelligence community. For this chapter, the term 'adverse action' refers to non-appealable and appealable personnel actions listed under chapter 75, which includes reductions in grade or pay, suspensions, and removals.[14] Adverse actions can be taken for reasons that will promote the efficiency of the service. Such actions include, but are not limited to, misconduct.

To address the fourth objective, we conducted a literature review to identify promising practices and lessons learned associated with employee misconduct in the federal sector. We interviewed officials from OPM, MSPB, the EEOC, and the Office of Special Counsel (OSC), to obtain their perspectives on responding to employee misconduct through alternative approaches. We also interviewed officials from the Environmental Protection Agency (EPA) to obtain their perspectives on recent efforts to better coordinate with EPA's Inspector General to address cases of employee misconduct. We obtained the perspectives of a panel of CHCOs from selected agencies through the CHCO Council. We also interviewed former human capital practitioners and other subject-matter experts with extensive experience working on employee misconduct issues identified through literature reviews and recommendations from other

[12] The Chief Human Capital Officers Act of 2002, enacted as part of the Homeland Security Act of 2002 (Pub. L. No. 107-296) required the heads of 24 Executive Departments and agencies to appoint or designate Chief Human Capital Officers (CHCOs). The CHCO Act also established a CHCO Council to advise and coordinate the activities of members' agencies on such matters as the modernization of human resources systems, improved quality of human resources information, and legislation affecting human resources operations and organizations.

[13] EHRI contains information on personnel actions and other data, and is the primary government-wide source for information on federal employees.

[14] 5 U.S.C. §§ 7502 and 7512.

experts. For further information on our scope and methodology, see appendix I.

We conducted this performance audit from July 2016 to July 2018 in accordance with generally accepted government auditing standards. Those standards require that we plan and perform the audit to obtain sufficient, appropriate evidence to provide a reasonable basis for our findings and conclusions based on our audit objectives. We independently assessed OPM and MSPB's data by determining data checks and quality control steps taken on entering and maintaining the data. We determined the data from both agencies to be sufficient and reliable for our purposes. We believe that the evidence obtained provides a reasonable basis for our findings and conclusions based on our audit objectives.

Background

Misconduct can occur in any workplace. When employee misconduct happens, an agency may incur a number of direct and indirect costs depending on how the agency chooses to address misconduct. For the agency, direct costs can mean potentially significant time and resource investments, including investigations, adversarial relationships between management and the employee, costs to the agency as a result of the misconduct committed (e.g., time and attendance or credit card fraud), and reduced employee engagement. The subject-matter experts we interviewed told us that, based on their experiences, the time it takes to address a case of misconduct may range from a couple of weeks to years. The time range depends on whether the employee appeals their case and other factors.

Agencies may also incur litigation expenses if an employee decides to appeal an adverse action. While there are costs to addressing misconduct, agencies also incur indirect costs when misconduct goes unaddressed in the workplace. These indirect costs include corrosive effects on other employees' morale, higher employee turnover, reduced productivity, and lower employee commitment to their work or agency. Indirect costs also include redirecting management's attention away from achieving the agency's mission.

Employee misconduct in the federal government is regulated by a well-developed body of statutes and regulations as well as decisions from MSPB and U.S. Court of Appeals for the Federal Circuit and Supreme Court. While there is no general definition of the term "employee misconduct" in a statute or government-wide regulation, Standards of Ethical Conduct are prescribed by the Office of Government Ethics at 5 CFR Part 2635 and agencies may also elaborate on types of misconduct in handbooks, tables of penalties (listings of some of the most common offenses with recommended ranges of penalties), and other internal guidance. There is a large body of law by MSPB addressing discipline for employee misconduct in the federal government that contains criteria of various forms of misconduct, such as, "insubordination," "excessive absence," and "misuse of government property." According to OPM officials, there are instances in law and regulation where types of misconduct are referenced concerning appointment into the competitive service. Chapter 73 (Suitability, Security, and Conduct) addresses certain types of misconduct of executive branch employees.

Generally speaking, an employee's violation of an agency's regulation or policy may cause the agency to take disciplinary or corrective action. Ultimately, if an agency needs to take an adverse action for inappropriate workplace behavior, it must do so "for such cause as will promote the efficiency of the service" as provided for in Title 5, Chapter 75.[15] OPM has also prescribed some regulations on employee responsibilities and conduct.[16] One of the nine Merit System Principles set forth by the Civil Service Reform Act of 1978 that govern the management of the federal workforce states that federal employees "should maintain high standards of integrity, conduct, and concern for the public interest."[17]

According to MSPB, when there is misconduct by a federal employee, management's goal should be to either persuade the employee to behave

---

[15] 5 U.S.C. § 7503(a) and 5 U.S.C. § 7513(a). Chapter 75 provides the procedures agencies must follow for actions that affect pay (e.g., suspensions and removals).

[16] 5 C.F.R. Part 735.

[17] 5 U.S.C. § 2301(b)(4).

properly or to remove the employee if the conduct is serious enough.[18] Moreover, OPM maintains that supervisors have a responsibility to set clear rules and expectations for employees in the workplace.[19] It is imperative that federal agencies manage their workforces effectively, which includes the effective use of discipline when addressing employee misconduct. In addition, employees may be disciplined for conduct that that they knew or should have known was unacceptable.[20] Similarly, federal executive branch employees have a responsibility to adhere to principles of ethical conduct and should avoid any actions that appear to violate the law or ethical standards.[21]

Overall, the objective of discipline is to deal with employees who are unwilling or unable to behave properly, and, where management deems it possible and appropriate, correct deficiencies in employee conduct. When management decides to take an action short of removal, discipline can deter misconduct and correct situations interfering with productivity. Conduct-based actions are important tools designed to aid supervisors in maintaining an efficient and orderly work environment.

---

[18] Merit Systems Protection Board, *Alternative Discipline: Creative Solutions for Agencies to Effectively Address Employee Misconduct* (Washington, D.C.: July 2008).

[19] For the purpose of this report, we apply MSPB's definition of supervisors as first-line supervisor who do not supervise other supervisors; typically those who are responsible for employees' performance appraisals and approval of their leave.

[20] As the Supreme Court observed in *Arnett v. Kennedy*, 416 U.S. 134 (1974), a case addressing the Constitutionality of the "efficiency of the service" in connection with an employee's removal for impugning his supervisors, "[I]t is not feasible or necessary for the Government to spell out in detail all that conduct which will result in retaliation.... The most conscientious of codes that define prohibited conduct of employees include `catchall' clauses prohibiting employee `misconduct,' `immorality,' or `conduct unbecoming.' We think it is inherent in the employment relationship as a matter of common sense if not [of] common law that [a Government] employee . . . cannot reasonably assert a right to keep his job while at the same time he inveighs against his superiors in public with intemperate and defamatory [cartoons].. . . [Dismissal in such circumstances does not come] as an unfair surprise ...." 416 U.S. at 158-164.

[21] 5 C.F.R. § 2635.101 (2011).

AGENCIES MUST FOLLOW STATUTORY AND REGULATORY PROCEDURES WHEN TAKING ADVERSE ACTIONS FOR EMPLOYEE MISCONDUCT UNDER CHAPTER 75 OF TITLE 5

Most agencies are required to adhere to formal, statutorily established guidelines under chapter 75 when taking adverse actions against an employee for misconduct. Chapter 75 of Title 5 includes two subsections that outline the requirements for (1) non-appealable adverse actions such as suspensions of 14 days or less[22] or (2) appealable adverse actions such as reductions in pay or grade, suspensions of more than 14 days, and removals (see Figure 1).[23] Subchapter I actions are covered by sections 5 U.S.C. 7501-7504 (Subsection I) and are referred to as "non-appealable actions," while Subchapter II actions covered by sections 5 U.S.C. 7511-7514 are referred to as "appealable actions" based on whether or not they can be appealed to the MSPB.[24]

According to a MSPB report, through the Civil Service Reform Act of 1978 (CSRA), "Congress sought to ensure that agencies could remove employees who engage in misconduct while protecting the civil service from the harmful effects of management acting for improper reasons, such as discrimination or retaliation for whistleblowing."[25] OPM regulations specify the process agencies must pursue to take adverse actions.[26] These regulations also specify the procedural and appeal rights to which employees facing adverse actions are entitled.[27] According to OPM officials, agency policies usually cover lesser disciplinary actions, such as oral and written reprimands, letters of warning, and letters of counseling. Employees may grieve these actions depending on the agency's administrative or negotiated grievance processes. According to OPM,

---

[22] 5 U.S.C. §§ 7501-7504.

[23] 5 U.S.C. §§ 7511-7514.

[24] 5 U.S.C. § 7701.

[25] MSPB, *What is Due Process in Federal Civil Service Employment?* (Washington, D.C.: May 2015).

[26] 5 C.F.R. §§ 752.203; 752.404; 752.604 (2012).

[27] See 5 CFR 752.405.

agencies may issue these actions without following the procedural requirements for adverse actions under 5 U.S.C. Chapter 75.

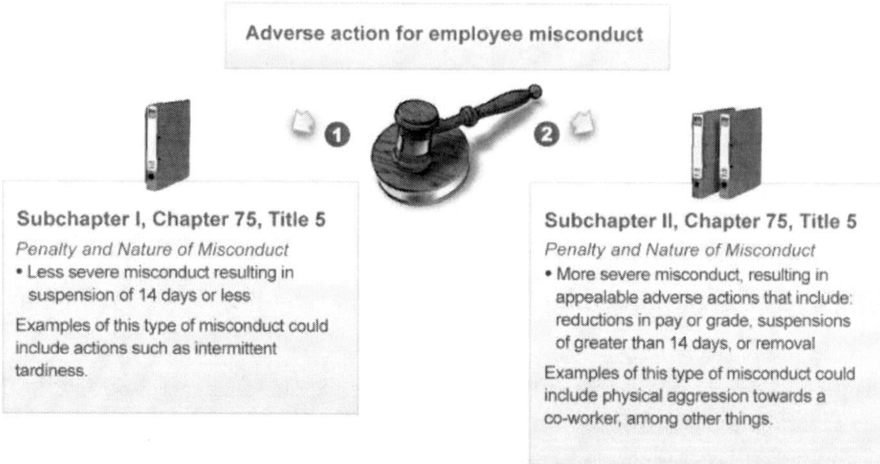

Source: GAO analysis of Chapter 75 of Title 5. | GAO-18-48.

Note: Outside of taking adverse actions under Subchapter I or II of chapter 75, Title 5 of the U.S. Code, agencies may take less severe actions to address misconduct, such as letters of counseling or oral admonishments, as well as official reprimands which are placed in an employee's official personnel file for a period of time.

Figure 1. Adverse Actions for Employee Misconduct under Subchapters I and II of Chapter 75, Title 5 of the U.S. Code.

The procedural rights due to employees subject to adverse actions covered by Chapter 75 are derived both from Chapter 75 and from the U.S. Constitution. In 1985, the U.S. Supreme Court held that tenured or post-probationary public employees who may be terminated only for cause have a constitutional property interest in continued employment and cannot be deprived of their jobs without due process of law. The process that was due in that case was notice of the proposed removal before it occurred and the opportunity to present reasons why the proposed action should not be taken.[28] Chapter 75 and OPM regulations promulgated thereunder establish additional procedural requirements extending to actions other than removal

[28] *Cleveland Bd. of Educ. v. Loudermill*, 470 U.S. 532, 546 (1985); see also *Gilbert v. Homar*, 520 U.S. 924 (1997).

that go beyond what the Due Process Clause of the U.S. Constitution would itself require under current precedent. For example, they require that employees be given advance notice of a suspension or a reduction in pay with the opportunity to respond in writing with supporting affidavits.

Subchapter I of Chapter 75 and Corresponding Regulations Describe Procedures for a Suspension of 14 Days or Less

An agency may take an adverse action under Subchapter I of Chapter 75 only for such cause as will promote the efficiency of the service. When proposing to suspend an employee for 14 days or less, an agency must give the employee advance written notice stating the reasons for the proposed suspension. The agency must also inform the employee of his or her right to review the material which is relied on to support the reasons for the action. The agency must give the employee a reasonable time (no less than 24 hours) to answer orally and in writing, to furnish affidavits and other documentary evidence in support of the answer, and to be represented by an attorney or other representative. Lastly, the agency is to give the employee a written decision with the specific reasons for the suspension on or before the effective date of the action.[29]

An employee may challenge a suspension of 14 days or less through an agency administrative grievance procedure, if applicable. If the employee is represented by a union with a collective bargaining agreement (CBA) with the agency that includes an applicable grievance procedure, the employee may challenge the suspension only under the CBA unless the employee is alleging that the suspension was discriminatory.[30] If the employee wishes to challenge the suspension as discriminatory or retaliatory under the EEO laws, the employee may file an EEO complaint with agency followed by a request for a hearing with the EEOC. The employee may also file a complaint with the OSC and then, if necessary, an Individual Right of Action appeal with the MSPB, to assert that the

[29] 5 U.S.C. § 7503(b)(1)-(4); 5 CFR § 752.203(a)-(e).
[30] 5 CFR § 751.103(f).

suspension was in retaliation for the employee's whistleblower activity. If the employee is represented by a union that has a collective bargaining agreement with the agency that includes an applicable negotiated grievance procedure, the employee may only file a grievance under the agency's collective bargaining agreement.[31]

Subchapter II of Chapter 75 Describes Steps Agencies Must Take to Address More Significant Cases of Employee Misconduct Than Those Addressed Under Subchapter I

Subchapter II of Chapter 75 addresses steps agencies must follow to take the four adverse actions listed below. These following actions are referred to as appealable adverse actions:

- suspensions longer than 14 days;
- reductions in grade;
- reductions in pay; and
- removals.[32]

An agency may take an adverse action under Subchapter II only for such cause as will promote the efficiency of the service. Subchapter II and OPM regulations contain more extensive procedural requirements for removals, reductions in pay or grade, and suspensions of over 14 days.[33] The employee is entitled to at least 30 days advanced written notice of the proposed action, unless there is reasonable cause to believe the employee has committed a crime for which a sentence of imprisonment may be imposed.[34] The notice must state the reasons for the action and inform the

[31] *Cleveland Bd. of Educ. v. Loudermill*, 470 U.S. 532, 546 (1985); see also *Gilbert v. Homar*, 520 U.S. 924 (1997).

[32] 5 U.S.C. § 7512(1)-(5). Subchapter II also applies to furloughs of 30 days or less. A "furlough" means the placing of an employee in a temporary status without duties and pay because of lack of work or funds or other non-disciplinary reasons.

[33] 5 U.S.C. §7511-7514 and 5 C.F.R. 752.404.

[34] See 5 CFR 752.404(d).

employee of their right to review the material on which the reasons stated in the notice are based.[35] Agencies typically provide the material supporting the proposal to the employee with the notice. The proposal is usually prepared by the employee's supervisor in consultation with human resources and, sometimes, the agency's legal staff.

The agency must give the employee a reasonable amount—no less than 7 days—of official time to review the supporting material, to prepare an answer orally and in writing, and to furnish affidavits and other documentary evidence in support of the answer.[36] According to OPM and MSPB officials, normally the agency designates an official other than the person who proposes the adverse action to review the employee's response and make the decision. The employee is entitled to be represented by an attorney or other representative, including a union steward if the employee is a bargaining unit member. The employee is entitled to a written decision on or before the effective date specifying the reasons for the decision and advising the employee of any appeal and grievance rights under 5 CFR § 752.405.[37]

An employee may challenge discipline under Subchapter II through an agency administrative grievance procedure, if applicable, or by filing a grievance under an applicable CBA. An employee may also appeal adverse actions covered by Subchapter II to the MSPB unless the employee first filed a grievance challenging the action under the CBA.[38] Accordingly, a large number of MSPB decisions address the elements and relative seriousness of various kinds of misconduct.

According to OPM, if the employee wishes to challenge an appealable adverse action under subchapter II as discriminatory or retaliatory under the EEO laws, the employee may file a "mixed case" EEO complaint with agency. The agency then issues a final agency decision that may be appealed to the MSPB. An employee affected by an appealable action (removal, suspension for more than 14 days, reduction in pay or grade) who believes that the action was motivated by prohibited discrimination,

---

[35] 5 CFR § 752.404(b)(1).

[36] 5 CFR § 752.404(c)(1).

[37] 5 C.F.R. § 752.404(c)-(e) and (g) (2012).

[38] 5 U.S.C. 7511-7512, 7514; 5 C.F.R. 752.405(a).

such as a person's race, color, religion, sex, national origin, age or disability, may also file a "mixed case" appeal directly with MSPB and raise the discrimination claim in that forum. The employee may seek review of MSPB's decision on the discrimination claim before the EEOC. If MSPB and EEOC disagree on the discrimination claim and MSPB does not defer to EEOC's view, then a special panel of the EEOC and MSPB will be convened to resolve the disagreement.[39]

When deciding an appropriate penalty for misconduct, agency officials are to make decisions on a case-by-case basis, taking into consideration all relevant circumstances. Deciding officials within the agency should consult the Douglas Factors –12 criteria developed by MSPB to guide such decisions (see appendix II). In the Douglas vs. Veterans Administration decision, MSPB found that a penalty will be sustained as long as "managerial judgment has been properly exercised within tolerable limits of reasonableness."[40] The list of Douglas Factors is not exhaustive. According to MSPB, weighing all relevant aggravating and mitigating factors and the totality of the circumstances is critical in any disciplinary case. The process agencies use to identify and address employee misconduct is illustrated in Figure 2.

Agencies may use progressive discipline to help determine which course of action to take when responding to misconduct. OPM officials define progressive discipline as the "imposition of the least serious disciplinary or adverse action applicable to correct the issue or misconduct with penalties imposed at an escalating level for subsequent offenses." The Douglas factors incorporate the concept of using a lesser penalty in appropriate circumstances. For instance, if an employee commits a first offense, the agency may choose to suspend the employee for 14 days or less. After that, the employee might learn from his or her mistake or

---

[39] 5 U.S.C. § 7702(d)(2)(A).

[40] MSPB, in its often cited decision, *Douglas vs. Veterans Administration,* 5 M.S.P.R. 280, 305 (1981), established criteria that supervisors must consider in determining an appropriate penalty to impose for an act of employee misconduct. The Board examined past court rulings and Civil Service Commission decisions regarding penalties and then summarized them into twelve (12) factors that it would look at to determine if a penalty was unreasonable. These factors are collectively known as the Douglas factors for the case that articulated them and they are still in use today.

correct the action, and not commit another offense, and therefore the agency will not discipline the employee again. However, the President has now prescribed that "supervisors and deciding officials should not be required to use progressive discipline," and that "the penalty for an instance of misconduct should be tailored to the facts and circumstance."[41] This will affect how agencies will determine appropriate penalties going forward.

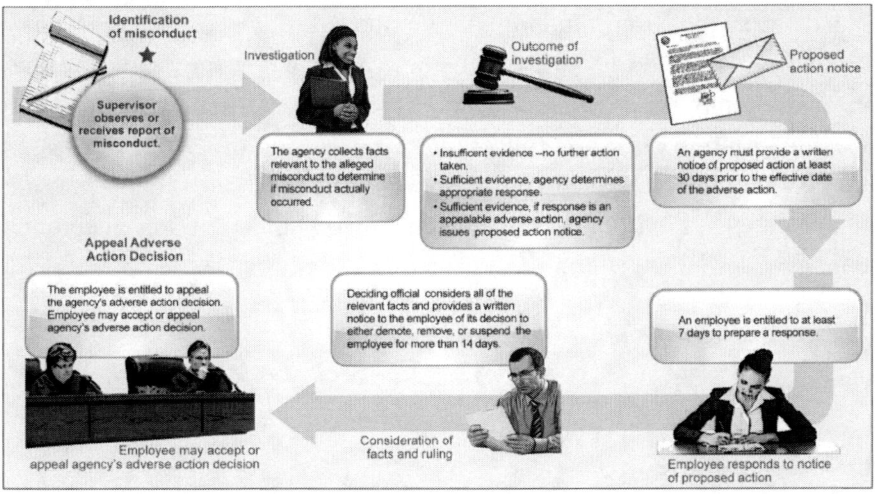

Source: GAO illustration of the process for addressing employee misconduct under Chapter 75 and 77 of Title 5. | GAO-18-48.

Figure 2. Process for Identifying and Addressing Employee Misconduct Using Chapter 75, Subchapter II and Chapter 77 of the U.S. Code.

Alternatively, if the employee commits the same offense a second time, the agency may choose to suspend the employee for longer, or impose stronger adverse actions, including removal. According to OPM officials, progressive discipline is not defined or required by civil service law, rules or regulations.

Chapter 75 provides that an employee with appeal rights who wants to contest an agency decision to remove, suspend for over 14 days, or reduce

---

[41] Section 2(b) of *Executive Order13839: Promoting Accountability and Streamlining Removal Procedures Consistent with Merit System Principles* (Washington, D.C.: May 25, 2018).

in pay or grade may appeal the agency's decision with MSPB.[42] If that employee is a member of a collective bargaining unit, the employee also has the option of pursuing a grievance under negotiated grievance procedures if the appeal has not been excluded from coverage by the collective bargaining agreement. The employee may pursue either option, but not both.[43] The employee may seek review of an arbitrator's decision before the U.S. Court of Appeals for the Federal Circuit. If the employee is challenging an adverse action within the jurisdiction of the MSPB and also alleged unlawful discrimination before the arbitrator or was prevented from doing so by the negotiated grievance procedure, the employee may appeal the arbitrator's decision to the MSPB. In addition, the union may appeal an arbitration award concerning a suspension of 14 days or less to the Federal Labor Relations Authority (FLRA) on behalf of the employee.[44] See Figure 3 for the collective bargaining unit appeals process for major disciplinary actions.

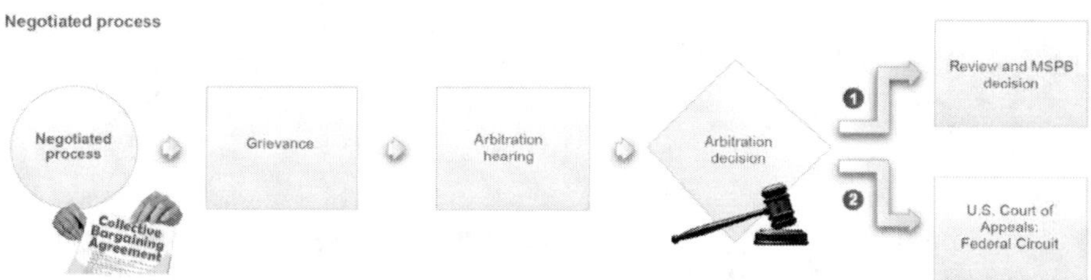

Source: GAO illustration of MSPB's flowchart of the Federal Employee Negotiated Review for major disciplinary actions. | GAO-18-48.

[1]An employee who claimed discrimination before an arbitrator or who was precluded by the negotiated grievance procedure from raising discrimination before the arbitrator may seek review of the arbitrator's decision before the MSPB. 5 U.S.C. § 7121(d); 5 C.F.R. § 1201.155(c).

[2]An employee may seek review of an arbitrator's decision before the U.S. Court of Appeals for the Federal Circuit. 5 U.S.C. §§ 7121(f), 7703.

Figure 3. Collective Bargaining Unit Appeals Process.

---

[42] 5 U.S.C. § 7513(d).
[43] 5 U.S.C. § 7121(e)(1); 5 C.F.R. § 752.405(b).
[44] See 5 C.F.R. Part 2425.

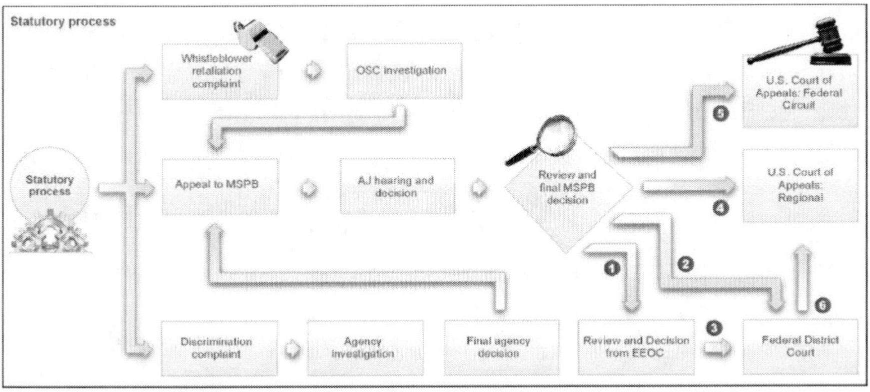

Source: GAO illustration of MSPB's flowchart of the Federal Employee Statutory Review Process for major disciplinary actions. | GAO-18-48.

[1] An employee who claimed discrimination in an appeal to the MSPB or in a grievance may seek review of the final MSPB decision before the EEOC. 5 U.S.C. § 7702(b).

[2] An employee who claimed discrimination in an appeal to the MSPB or in a grievance may file a civil action in federal district court. 5 U.S.C. § 7703(b)(2).

[3] An employee may file a civil action in federal district court following a final decision by the EEOC. See authorities cited in 5 U.S.C. § 7702(a) & (b)(5)(A) and 29 C.F.R. §§ 1614.407 & 1614.408.

[4] An employee who filed a whistleblower complaint may obtain judicial review of a final decision of the MSPB before the U.S. Court of Appeals for the appropriate regional circuit 5 U.S.C. § 7703(b)(1)(B).

[5] An employee may obtain judicial review of a final decision of the MSPB before the U.S. Court of Appeals for the Federal Circuit. Such review may include a claim of whistleblower retaliation but may not include a claim of discrimination. 5 U.S.C. § 7703(b)(1)(A) & (B); Kloeckner v. Solis, 133 S. Ct. 596 (2012).

[6] An employee who pursued a civil action in federal district court may seek judicial review of the district court in the U.S. Court of Appeals for the relevant regional circuit. 28 U.S.C. § 1291.

Figure 4. Statutory Appeals Process for Appealable Adverse Actions.

Employees may use several avenues if they elect to appeal adverse actions through the statutory appeals process for such actions (removal, suspension of more than 14 days, and reduction in grade or pay).[45] If the

[45] See 5 CFR 752.405.

employee believes the disciplinary action was motivated by unlawful discrimination, he or she may file a discrimination complaint with the agency or file an appeal directly with MSPB. If the employee believes the disciplinary action was taken in retaliation for whistleblowing, he or she may choose to file a whistleblower retaliation complaint before deciding to appeal to the MSPB. If the employee or agency does not agree with the decision rendered by a MSPB administrative judge (AJ), he or she may seek review before the full MSPB. See Figure 4 for statutory appeals process.

Initial Appeals to MSPB Take Around 100 Days to Resolve

We analyzed MSPB's data and found initial appeals at MSPB generally take from 63 to152 days to render a decision.[46] MSPB has a policy goal of resolving cases by an administrative judge on or before 120 days after the filing of the appeal. An employee or agency can appeal an initial MSPB decision in a process called petition for review (PFR). PFR cases are reviewed by the full MSPB, and range from an additional 99 to 251 days, based on our analysis of the MSPB's data. The time that it takes to resolve cases at MSPB is consistent for demotions, suspensions of greater than 14 days, and removals. According to MSPB officials, the system is designed to require an individual to choose a path of review to the exclusion of other paths. Depending on the claims raised, there may be multiple levels of review of a single action before multiple fora. However, there is only one hearing at the administrative level; therefore, the timeline to resolve an adverse action appeal can be longer than the initial appeal and PFR.

According to MSPB officials, the selected CHCOs, and the subject-matter experts we interviewed, agencies most often make the following errors which may cause MSPB to reverse the adverse action decision[47]:

---

[46] Range is based on the 25th and 75th percentiles.

[47] Penalties imposed may vary depending on how agencies label different types of misconduct.

- *Failure to follow procedures by agency:* MSPB may overturn an adverse action decision if the agency did not adhere to the processes set out in statute and regulation. This most often means that the agency did not give the employee a chance to respond to the adverse action charge or did not notify them of their rights to an attorney.
- *Failure to follow procedures by deciding official:* An action may be vulnerable to a modification or reversal upon appeal if the deciding official did not fulfill their role appropriately in weighing the evidence through a Douglas Factors analysis.
- *Ex parte communications:* A challenge may be overturned if a deciding official gave consideration to any issue not in the proposal letter.
- *Incorrect labeling (or charge):* Nothing in law or regulation requires an agency to attach a label to a charge of misconduct. However, if labels are used, they must be proven. An example used by MSPB provides that if an agency uses the label of "theft" as its charge, then the agency must prove that the employee "intended to permanently deprive the owner of possession" of the item in question.[48] Experts told us that MSPB requires agencies to prove all legal aspects of a misconduct label.

Federal courts have held that it is impermissible to allow the official who makes the final decision in a removal proceeding to rely on aggravating factors regarding either the alleged offense or the proposed penalty that were not contained in the notice, and to which the employee did not have an opportunity to respond.[49] MSPB is bound by this precedent.

---

[48] *King v. Nazelrod*, 43 F.3d 663, 666-67 (Fed. Cir. 1994).

[49] Aggravating factors could include prior disciplinary actions, poor work record, lack of rehabilitative potential, etc. See *Ward v. U.S. Postal Service*, 634 F.3d 1274 (Fed. Cir. 2011); *Stone v. F.D.I.C.*, 179 F.3d 1368 (Fed. Cir. 1999).

Alternative Disciplines Can Help Agencies in Determining the Appropriate Response to Misconduct but Several Factors Affect How Agencies Respond

Alternative discipline is an approach to address misconduct that is available to agencies in lieu of traditional penalties (e.g., letters of reprimand and suspensions of 14 days or less). According to MSPB, agencies may choose to offer alternative discipline at any stage of the disciplinary process. OPM officials said alternative disciplines tend to be more focused on taking a corrective or remedial response rather than punitive actions against an employee. In a report on alternative discipline, MSPB states that alternative discipline can take many forms and is an effort undertaken by an employer to address employee misconduct using a method other than traditional discipline.[50]

As an alternative discipline approach, it is recommended by MSPB that agencies may consider entering into an agreement with an employee. In general, such an approach involves a legally binding written agreement between the employee and the agency addressing an act of misconduct. If the employee violates the agreement, the agency will proceed with additional or more serious forms of discipline, up to and including removal. MSPB also recommends that managers and human resources personnel consult with legal counsel when drafting and implementing an alternative discipline agreement that requires the employee's consent, adding that it is extremely important for agreements to meet certain legal requirements to form a valid agreement.

We compiled a non-exhaustive list of alternative discipline based on a literature review and interviews. Subject-matter experts, including the panel of CHCOs, reviewed this list and they cited benefits and drawbacks to some of the approaches (see Table 1).

---

[50] *Alternative Discipline: Creative Solutions for Agencies to Effectively Address Employee Misconduct (July 2008).*

Table 1. List of Alternative Discipline Approaches and Details of the Approach

Alternative discipline	**Details of approach**
Additional training	Employee attends training designated by the agency that addresses specific acts of misconduct, such as time and attendance infractions.
Involvement in process improvement	The employee can develop or lead training in proper employee behavior for that specific area of misconduct in which the employee engaged.
Rescinding telework or alternate work schedule agreement	Preventing a telework eligible employee from teleworking or eliminating an alternative work schedule as part of the disciplinary action.
Counseling	Employee attends counseling for problems that are causing him or her to act inappropriately; types of counseling may include anger management, drug and alcohol treatment, and more. If the employee agrees to counseling, the penalty can be held in abeyance or laid out in the last chance agreement.
Community service	Employee agrees to complete a specific number of community service hours equal to the amount of time spent on a suspension.
Alternative dispute resolution (ADR)	A third-party mediator is used to settle disputes between two or more parties. ADR is more cost-effective than going to litigation before an adjudicating board.
Last chance agreements	In the event of an ongoing disciplinary history or egregious first-offense misconduct, an agency can hold a removal action in abeyance by entering into a written agreement with an employee. The agreement should state that the employee will not repeat the act of misconduct or any similar act of misconduct and should remain in effect for a set period of time. If the employee engages in the same or similar misconduct again, he or she can be removed without further procedural rights or appeal rights.
Abeyance agreements	Management holds a disciplinary action in abeyance, meaning that the discipline is not immediately enacted, for a set period of time or only imposes a portion of the disciplinary action, e.g., a suspension.

Table 1. (Continued)

Alternative discipline	Details of approach
	This agreement is also in writing. If the employee commits another act of misconduct during this set period of time, then management enacts discipline. If an employee commits an infraction that he or she is unaware is misconduct, then an abeyance agreement might be appropriate. The employee agrees to waive grievance or appeal rights if applicable.
Paper suspension agreement	An employee receives a suspension on paper, which is documented on their permanent or temporary record, but the employee is not actually suspended on a non-pay status. The paper agreement states that the paper suspension is counted as progressive discipline and is of the same weight as a traditional suspension if there is additional misconduct. Employee agrees to waive appeal rights.
Clean-slate agreements	In lieu of disciplinary action or removal, the employee voluntarily agrees to separate, and waives appeal rights. In return, the agency agrees that the employee will have a clean slate, meaning the misconduct will not be permanently recorded in his or her personnel file. The employee must be honest about the reason prompting the separation in applying for future employment, filling out forms for background investigations, or responding to investigators' inquiries in connection with background investigations. Agencies must follow ethical requirements with respect to providing references.
Financial agreements	If an agency proposes an adverse action and the employee in question responds with a complaint, the agency is authorized to buy out the employee, meaning the agency will pay the employee a sum of money to separate from the agency. Employee agrees to waive appeal rights as part of the buyout process.
Leave donation in lieu of formal discipline	An employee voluntarily agrees to donate annual leave to the agency's leave bank in lieu of disciplinary action.
Reassignment of employee or manager	Employee or manager is reassigned to a new team or different part of the agency. Must be mutually agreed upon and consistent with any collective bargaining agreements that the employee is under.

Alternative discipline	Details of approach
Financial restitution for cases involving misuse of government property	This would require a full investigation and the agency would specify how the employee would be required to make restitution.
Employee issues a public apology to affected individuals and or agency	Employee issues a public apology to the agency or individuals affected by the misconduct.

Source: GAO summary of feedback provided by agency officials and subject-matter experts. | GAO-18-48.

MSPB noted in its 2008 report that the specific alternative discipline approach an agency decides to use should be based on the nature and severity of the misconduct. According to OPM officials, alternative discipline approaches are not appropriate for egregious acts of misconduct or when the employee is remorseless but rather lower level offenses where an employee may show remorse for the misconduct and demonstrate that she or he can be rehabilitated. Egregious acts of misconduct may involve discrimination, reprisal or retaliation, or sexual harassment. Alternative discipline approaches are also not appropriate when the employee's continued presence in the workplace would pose a threat to the employee or others.

On a case by case basis, an agency may decide to provide counseling or additional training as appropriate, depending on the facts and circumstances that address specific acts of minor misconduct. Additionally, agencies have flexibility in using alternative discipline as a final effort before taking formal action such as suspension or removal. However, in its 2008 report, MSPB found that managers had applied alternative discipline approaches ineffectively, resulting in further inefficiencies in the civil service. Specifically, MSPB recommended that managers and human resources personnel consult with legal counsel when drafting and implementing an alternative discipline agreement that requires the employee's consent, adding that it is extremely important for agreements to meet certain legal requirements to form a valid agreement. CHCOs and subject-matter experts said managers and supervisors should coordinate internally with human resources staff, employee relations, and legal counsel when assessing whether an alternative discipline approach would result in correcting improper behavior and ultimately improve their workforce.

Some subject-matter experts we interviewed expressed concern that workforces would view alternative discipline measures as providing opportunities for employees to avoid accountability or encouraging similar negative behaviors from coworkers rather than penalizing the employee more stringently through a formal adverse action process. These subject-matter experts identified community service, buy-outs, involvement in process improvements, and clean-slate agreements as approaches that had this kind of effect. Additionally, some subject-matter experts told us that alternative discipline approaches such as community service and paper suspension agreements could have the unintended effect of benefiting the employee being disciplined. For example, community service may allow the employee to serve the alternative discipline during their scheduled duty time instead of performing their regularly assigned duties. This may require the employee's co-workers to take on additional work while the employee serves the alternative discipline. Additionally, while a paper suspension limits interruption to work production, it also allows the employee to work in a pay status while carrying out the suspension. According to feedback we received from the CHCOs, some of these alternative discipline approaches were used more often than others and some approaches were more effective at addressing employee misconduct. We did not evaluate how often or the extent to which any of these approaches are used at agencies, nor did we consider the propriety or legality of these approaches.

According to the CHCOs and subject-matter experts, agency managers and supervisors may be able to effectively resolve employee misconduct cases through the use of alternative approaches, which can shorten the timeline and simplify the adverse action process in a manner that has the most potential to prevent additional harm to the workplace and avoid the potentially high costs of litigating a misconduct case.

Several Factors Can Affect Whether and How an Agency Addresses Employee Misconduct

Current and former agency officials and subject-matter experts we interviewed told us in interviews that several factors can affect whether and how an agency responds to misconduct. Both agency officials and subject-matter experts told us that supervisors may not report misconduct due to fear that an employee could counter with their own complaint. Several CHCOs and subject-matter experts told us that an agency's approach to dealing with misconduct can influence how first-line supervisors act. In a recently released MSPB publication that highlighted selected results of its 2016 Merit Principle survey of managers and supervisors about challenges to addressing employee misconduct, 80 percent of managers and supervisors agree to some extent or a great extent that their agency's culture poses a challenge when attempting to remove an employee for serious misconduct.[51] Additionally, MSPB's report provided the perspectives of managers and supervisors regarding the factors that affect how agencies address misconduct, including

- 77 percent of managers/supervisors agree to some extent or a great extent that they do not feel supported by their agencies' senior leadership in their actions to remove an employee for serious misconduct.
- 88 percent of managers/supervisors somewhat or strongly agree that some supervisors do not manage their employees' conduct because the supervisors want to avoid conflict.
- 64 percent of managers/supervisors agree to some extent or a great extent that they do not fully understand the process to remove an employee for misconduct.

---

[51] The MSPB 2016 Merit Principles survey is a government wide survey of federal employees and supervisors that covered a variety of workforce issues, prohibited personnel practices, and selected aspects of their work experience and work environment. Merit Systems Protection Board, *Addressing Misconduct in the Federal Civil Service: Management Perspectives* (Washington, D.C.: December 2016).

MSPB's 2016 survey findings were consistent with what agency officials and subject-matter experts told us during interviews.

Removal Actions for Misconduct Taken under Chapter 75 Are Relatively Rare

Our analysis of OPM data from fiscal year 2006 to 2016 shows that, on average, agencies disciplined approximately 17,000 or less than 1 percent of the federal workforce per year under Subchapter II of Chapter 75.[52] The number of employees who separate from the federal workforce for misconduct under alternative means, such as settlements, is not known and would not be recorded as misconduct in OPM's EHRI database, according to agency officials and experts. Many of the CHCOs and subject-matter experts we interviewed told us that while data around such cases are not collected government-wide, they believe internal resolutions using alternative approaches to address misconduct occur frequently.

Trends in Misconduct Removals Are Associated with Fluctuations in Probationary Employee Numbers

According to EHRI data, as the number of probationary employees fluctuated over time, the number of terminations generally followed the same trend. One of the likely reasons for this fluctuation is that probationary employees are more likely to be terminated than career employees who are no longer in a probationary status because probationary employees are not yet subject to the Chapter 75 process protection afforded career employees.

[52] For our analysis of EHRI data, the observational unit is an adverse action, such as removal, suspension, or demotion, rather than an individual employee because employees may have multiple adverse actions recorded in EHRI.

Similar to addressing performance issues, it is generally easier to terminate employees for misconduct during the probationary period. As we previously reported, the probationary period is an important management tool to evaluate the conduct and performance of an employee and should be treated as the last step in the hiring process. According to OPM, appropriate actions taken within the probationary period are the best way to avoid long-term problems.

The Most Widely Used Form of Formal Discipline for Misconduct Is Suspension; Approximately One-Fourth of Suspended Employees Have Multiple Suspensions

Our data analysis of personnel actions against employees for misconduct shows that the most common form of discipline is suspension. In 2016, agencies made 10,249 suspensions, 7,411 removals, and 114 demotions for misconduct (the numbers refer to the number of adverse actions that agencies made in 2016, not the number of employees that received adverse actions; one employee can be suspended multiple times, and each suspension is recorded as a separate personnel action in the employee's SF-50).[53] The data we analyzed indicated that approximately one-fourth of suspended employees have multiple suspensions. According to OPM officials, third parties such as the MSPB will review whether disciplinary actions are taken "only for such cause as will promote the efficiency of the service" which includes the assessment of the relevant Douglas factors.

Figure 5 shows how many suspensions, demotions, and removals took place from fiscal years 2006 to 2016 according to EHRI data.

---

[53] 2016 is the most recent year for which data were available.

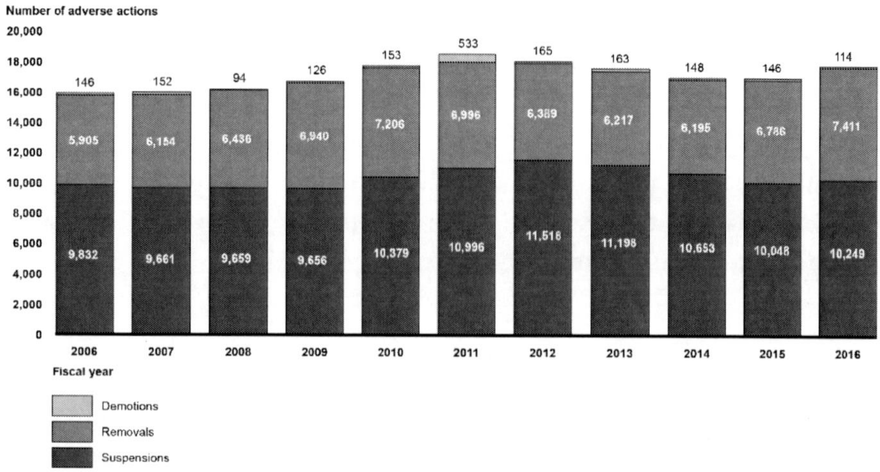

Source: GAO analysis of Enterprise Human Resource Integration data. | GAO-18-48.

Note: For our analysis of EHRI data, the observational unit is an adverse action, such as removal, suspension, or demotion, rather than the individual employee because employees may have multiple adverse actions recorded in EHRI.

Figure 5. Number of Suspensions, Demotions, and Removals Per Year, Fiscal Years 2006 to 2016.

Better Data on Employee Misconduct Could Strengthen OPM's Oversight and Provide Clarity to Agencies Regarding How to Address Misconduct

OPM collects data on personnel actions reported by most agencies and stores this information in the EHRI database, but these data could be improved to provide OPM with better information to help agencies address misconduct.[54] Because not all misconduct data are entered into the database, the data presented in this chapter do not represent the entirety of employee misconduct instances that occur in the federal government.[55]

---

[54] Our analysis of the data does not represent a minimum or maximum of misconduct cases. Rather, it represents the number of adverse disciplinary actions taken under Subchapters I and II of Chapter 75 that agencies took and reported to OPM through the EHRI database.

[55] EHRI captures actions that generate an SF-50 such as suspensions, removals, reduction in grade, and reduction in pay. It does not capture lesser penalties for which OPM does not regulate.

Personnel actions in the EHRI database originate from data that agencies send to OPM through the Standard Form 50 (SF-50), a form that documents personnel actions. OPM officials told us that lesser disciplinary actions such as a letter of reprimand are not documented by an SF-50. Without maintaining comprehensive data regarding the extent and nature of misconduct in the federal government, OPM risks missing opportunities to provide agencies with guidance and other tools, such as targeted training to help agencies better address cases of misconduct. Indeed, better data could help OPM and agencies identify systemic misconduct issues, such as misuse of government property or physical aggression toward a co-worker, as well as emerging problems that benefit from early detection and/or more comprehensive approaches.

It should be noted that for the codes that indicate performance or misconduct as the underlying cause for the adverse action, it is not possible to make a clear distinction between whether the action was specifically related to misconduct, performance, or a mix of the two. Therefore, some cases include a mix of employee poor performance and misconduct. OPM officials said they do not have a sense of how frequently agencies use these (and other non-specific) nature of action (NOA) codes for misconduct-related actions. According to OPM officials, by establishing rules in terms of improving the efficiency of the service and the types of actions that will require specific procedures, Congress provided managers with maximum flexibility to pursue adverse actions whenever it would promote the efficiency of the service, whether the underlying impetus was a conduct issue or a failure to perform.

OPM officials told us the Guide to Processing Personnel Actions directs agencies to indicate the nature of personnel actions in the EHRI database through the NOA codes. These codes indicate the employee type, the nature of the personnel action to be recorded in EHRI, as well as the underlying cause (e.g., conduct or performance) for the personnel action. OPM performs validity checks on the NOA codes and legal authorities to assure the agencies are compliant with OPM reporting requirements. OPM also periodically reviews agencies' use of NOA codes and legal authorities in general.

The EHRI database does not collect or store the specific type of misconduct—only that the personnel action belongs in the misconduct category. Several CHCOs and subject-matter experts who we interviewed agreed this flexibility is helpful to agencies. For example, officials said that while common types of misconduct exist, such as time-and-attendance infractions, many unique types of misconduct cannot be placed into easily identifiable categories. The officials added that it would be easy for agencies to mislabel misconduct. For instance, OPM officials said that disobeying an agency's policy or rules could manifest itself in many different ways.

Moreover, we found inconsistencies in the data OPM provided. For example, during this review, we initially used stored EHRI data from previous audits for fiscal years 2006 to 2014. We used NOA codes provided by OPM officials to analyze employee misconduct data in the executive branch. When we compared the results of our data analysis for this period to the data OPM provided for the same period, we found their data identified approximately 500 more adverse actions per year. Though we consulted with OPM, we were unable to resolve these differences. OPM officials noted that agencies submit data on a rolling basis and may later correct it, and, some SF-50 forms are filed after the fiscal year ends, so our stored data may not include these actions.

According to OPM officials, agencies generally have day-to-day oversight for determining use of NOA codes and legal authorities. Agencies are required to report a valid NOA code and legal authorities that is found in OPM's Guide to Data Standards. Guidance to agencies for classifying misconduct into the correct nature of action codes is provided in *The Guide to Processing Personnel Actions*. Although OPM verifies that agencies provide valid NOA codes in their data, they assert that agencies have responsibility for determining which NOA codes to use for each personnel action based on OPM documentation.

As we noted in a 2017 report on federal human resources data, OPM developed EHRI to (1) provide for comprehensive knowledge management and workforce analysis, forecasting, and reporting to further strategic management of human capital across the executive branch; (2) facilitate the electronic exchange of standardized human resources data within and across agencies and systems and the associated benefits and cost savings; and (3) provide unification and consistency in human capital data across the executive branch.[56]

An important part of OPM's role is to support federal agencies' human capital management activities, which includes ensuring that agencies have the data needed to make staffing and resource decisions to support their missions. EHRI data are essential to government-wide human resource management and evaluation of federal employment policies, practices, training, and costs. The ability to capitalize on this information is dependent, in part, on the reliability and usefulness of the collected data. According to Federal Internal Control Standards, management is to obtain relevant data from reliable internal and external sources in a timely manner based on the identified information requirements.[57]

More specific guidance from OPM to agencies on which NOA codes to use for misconduct cases will increase confidence in the data, without requiring practitioners to capture and tabulate the type of misconduct. More importantly, enhanced data on the extent and nature of misconduct will improve OPM's oversight ability and agencies' ability to target management training and identify specific trends in misconduct.

Most MSPB Appeals Are Resolved by the Parties, Which Benefits Both the Agency and Appellant, According to Officials

Our analysis of MSPB data found that the most frequent appeal outcome is a settlement. MSPB said settlements often benefit both the agency and the appellant because they manage risk. The officials said that

---

[56] GAO, *Federal Human Resources Data: OPM Should Improve the Availability and Reliability of Payroll Data to Support Accountability and Workforce Analytics,* GAO-17-127 (Washington, D.C.: Oct.7, 2016).

[57] Internal controls help agency program managers achieve desired results and provide reasonable assurance that program objectives are being achieved through, among other things, effective and efficient use of agency resources. GAO, *Standards for Internal Control in the Federal Government*, GAO-14-704G (Washington, D.C.: September 2014); effective in fiscal year 2016.

when entering an adverse action appeal with MSPB, both the agency and the appellant face a risk: the agency is at risk of spending time and money for litigation only to, in some cases, have its decision overturned; the appellant is at risk of being removed from his or her position with a permanent mark on his or her record, which may make finding another job difficult. To avoid these outcomes for both parties, an agency may offer the employee a variety of settlement options to incentivize them to willingly leave. Settlement options may include, but are not limited to

- back-pay for the time that the employee was out of work, but still litigating the appeal; and
- paying the employee's attorney fees.

OPM notes, on the other hand, that the MSPB's data might not always reflect voluntary settlements. According to OPM officials, the MSPB has a large caseload and typically strives to induce the agency to settle. OPM officials noted that the pressure to settle cases regardless of merit after the agency has made the determination that discipline is necessary and has gone through the procedure to carry it out may be one of the most significant deterrents to dealing with misconduct or performance under Chapter 75.

Figure 6 shows the number of MSPB appeals that were filed from fiscal years 2006 to 2016 that were affirmed, reversed, settled, or dismissed.

We also analyzed data from MSPB's database of appeals cases.[58] MSPB hears appeals from those adverse actions that Congress made appealable under Subchapter II of chapter 75, including suspensions of greater than 14-days, demotions, and removals. These actions can be taken for performance problems as well as misconduct under Chapter 75—MSPB does not differentiate between performance and misconduct in its database. Rather, the agency categorizes its cases by legal authority. Therefore, similar to OPM's EHRI data, any analysis with MSPB's data

[58] When we refer to adverse actions regarding MSPB data, we are referring to suspensions longer than 14 days, demotions, and removals.

may include performance appeals as well as misconduct appeals under Chapter 75.

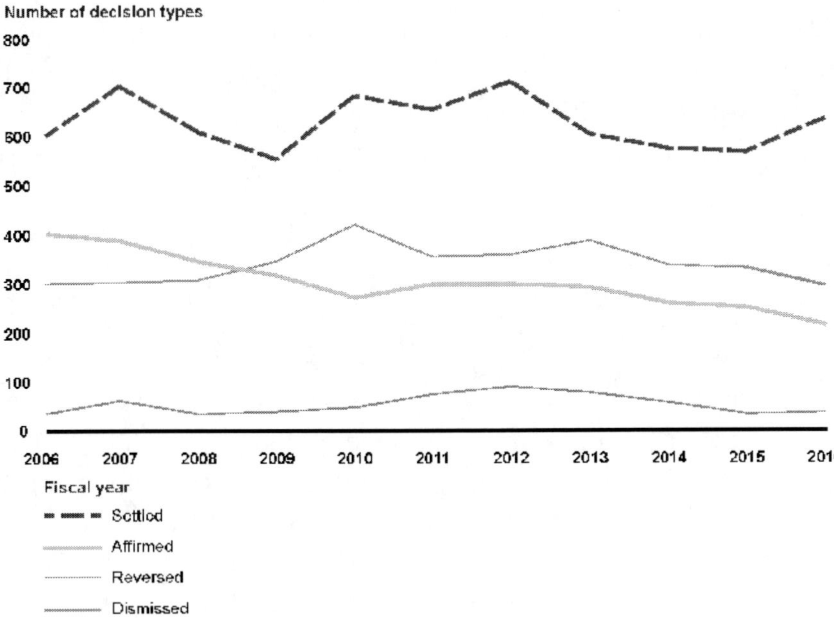

Source: GAO analysis of Merit Systems Protection Board appeals data. | GAO-18-48.

Note: For our analysis of MSPB data, the observational unit is an appeal action such as affirmed, reversed, settled, or dismissed, rather than the unit being an individual employee because employees may appeal cases several times to MSPB. We omitted "mitigated" and "other" as observational units because very few cases are mitigated by MSPB and the other category covered a wide range of outcomes.

Figure 6. Outcomes of Merit Systems Protection Board Appeals, Fiscal Years 2006 to 2016.

KEY STEPS AGENCIES CAN TAKE TO BETTER PREVENT AND ADDRESS EMPLOYEE MISCONDUCT

On the basis of our literature review, as well as interviews with CHCOs and subject-matter experts, we identified key promising practices and lessons learned that can help agencies better prevent and address employee misconduct. These key practices include tables of penalties, engaging employees, making full use of the probationary periods, and maintaining effective lines of communication and collaboration between the human resources office staff, line-level management, and agencies' legal counsel. Going forward, it will be important for OPM and agencies, in concert with the CHCO Council to examine each of these practices and lessons learned, refine, as appropriate, and share how best to implement these practices.

Tables of Penalties Can Help Guide Responses to Misconduct

We found that tables of penalties—a list of recommended disciplinary actions for various types of misconduct—though not required by statute, case law, or OPM regulations, nor used by all agencies, can help ensure the appropriateness and consistency of a penalty in relation to an infraction. Further, tables of penalties can help ensure the disciplinary process is aligned with merit principles because they make the process more transparent, reduce arbitrary or capricious penalties, and provide guidance to supervisors.

According to the panel of CHCOs and the subject-matter experts we interviewed, a table of penalties may also provide information on the period over which offenses are cumulative, for purposes of assessing progressively stronger penalties. The officials described tables of penalties as a listing of common infractions committed most frequently by agency employees, along with a suggested range of penalties for first, second and third offenses; however, the range of penalties should not be too broad, and the penalties should be progressive, meaning that they increase in harshness with each subsequent offense committed by the employee. The CHCOs and subject-matter experts said a table of penalties should also provide sufficient flexibility in the penalty range (e.g., 1-day to 5-day suspensions for a first offense) to consider mitigating and aggravating factors when considering discipline for misconduct. OPM officials stated that where an agency elects to have a table of penalties, it should serve as a guide in addressing misconduct, noting that it does not serve as a substitute for management's judgment. According to OPM officials, management must take into account the applicable Douglas Factors, and must consider other appropriate circumstances not covered by the Douglas Factors. Neither OPM nor MSPB provide any written guidance to agencies in developing their tables of penalties. However, OPM officials told us their agency is available to provide assistance upon request to agencies that elect to use a table of penalties.

Views on the usefulness of the tables of penalties were mixed among agency officials and subject-matter experts. On the one hand, some agency officials and other subject-matter experts told us the table of penalties can assist agencies in determining an appropriate penalty and ensure consistency of penalty selection from case to case. For that reason, they said the tables can also help ensure the action taken is legally defensible based on past similar cases. MSPB officials told us that they believe table of penalties, which rely on the Douglas Factors, can help human capital practitioners when making decisions about employee misconduct cases. On the other hand, several subject-matter experts and agency officials, including OPM, indicated that table of penalties tend to be too broad in the range of penalties for individual offenses, which they said ultimately limited their usefulness in the decision-making process.

OPM officials said their agency does not use a table of penalties nor does it support encouraging agencies to establish tables of penalties. According to OPM officials, where table of penalties exist, they are established at an agency's discretion and not under OPM's auspices. OPM officials believe agencies have the ability to address misconduct appropriately without a table of penalties and with sufficient flexibility to determine the appropriate penalty for each instance of misconduct. Further, OPM officials said that agencies that adopt a table of penalties will be required to consider its table of penalties, if applicable, as part of the MSPB's Douglas Factors analysis which, in their view, imposes an additional condition on the agency's ability to defend its actions. Finally, OPM said there is no substitute for management judgment and that tables of penalties should not be applied so inflexibly as to impair consideration of other factors relevant to the individual case.

In short, tables of penalties, if drafted at an appropriate level of detail and used in conjunction with the Douglas Factors and other case-specific forms of discretion, could provide agencies with reasonable assurance that similar cases of misconduct are addressed with similar penalties as appropriate, and can reduce the risk of inconsistently and potentially unfairly applying remedial measures.

Set Clear Expectations and Engage Employees

Agency officials and subject-matter experts told us that having effective agency policies and programs that set clear expectations around behavior and that engage employees may help reduce the number of misconduct incidents that occur. These policies and programs may also mitigate the damage when an incident does occur. Several subject-matter experts said that agencies should set formal expectations early and reinforce these expectations throughout an employee's career. To this point, as we discussed in our 2015 report on addressing substandard employee performance, when addressing misconduct agencies should help managers and supervisors take appropriate action if misconduct occurs during an employee's probationary period.[59]

Some subject-matter experts indicated that agencies may also consider conducting more thorough job screening and hiring processes which could help determine if the individual is a good fit for their agency. Specifically, the subject-matter experts mentioned that agencies should take a closer look at a prospective employee's work history and carefully check references. We also learned from our interviews that, as a deterrent, agencies must clearly communicate that an employee will be held accountable for any acts of misconduct.

---

[59] GAO, *Federal Workforce: Improved Supervision and Better Use of Probationary Periods Are Needed to Address Substandard Employee Performance*, GAO-15-191 (Washington, D.C.: Feb. 6, 2015).

According to OPM, agencies can mitigate the risks of these difficulties by establishing a well-trained, experienced, and empowered employee and labor relations staff. OPM said these individuals play a crucial role in educating supervisors and managers in taking appropriate and sustainable disciplinary actions. CHCOs and subject-matter experts provided a number of key promising practices that an agency can use to mitigate and address employee misconduct, including:

- Demonstrating positive conduct at the agency's senior leadership (tone at the top): Through policies and their own individual actions, senior leaders must exhibit positive workplace behavior as an example to agency employees.
- Maintaining a good workplace atmosphere: Agencies should take steps to monitor workforce morale and initiate programs that encourage respect and community.
- Engaging employees by connecting them directly to the agency's mission: Employees should have a sense of purpose and commitment toward their employer and its mission which can lead to better organizational performance.
- Making full use of the probationary period for employees: Supervisors should use probationary periods as an opportunity to evaluate an employee's performance and conduct to determine if an appointment to the civil service should become final.
- Setting and communicating clear rules and expectations regarding employee conduct: Agencies should set expectations about appropriate conduct in the workplace and communicate consequences of inappropriate conduct at the earliest possible time after on-boarding an employee.
- Assuring that employees conform to any applicable standards of conduct: Supervisors and managers, with the support of their agencies' leadership and human resources staff, should train and monitor employee compliance with its stated conduct policies.
- Maintaining effective lines of communication and collaboration with the human resources office staff, line-level management, and

agencies' legal counsel: Agencies should establish clear lines of communication across relevant offices to ensure misconduct cases are addressed effectively and consistently.

- Conducting on-going training for supervisors and holding them accountable for addressing misconduct in a timely manner when it occurs: Supervisors should be trained in identifying employee misconduct cases and knowledgeable about the process for addressing such cases.

More Effective Training Could Help Supervisors Identify and Deal with Misconduct

MSPB and OPM officials as well as subject-matter experts said human resources staff and line-level supervisors and managers would benefit from additional training in how to address employee misconduct. The subject-matter experts told us that managers do not receive sufficient training in how to identify and subsequently deal with misconduct in the workplace. Specifically, subject-matter experts told us that many supervisors and managers do not understand the requirements needed to remove an employee for misconduct, including misconceptions about the standard of proof required. Many subject-matter experts repeated observations MSPB made in its 2008 report that without sufficient training managers and supervisors may find it difficult to engage in challenging one-on-one conversations with an employee about misconduct.

Agency officials and subject-matter experts also told us that supervisory training varies by agency. Our subject-matter experts said some agencies are more structured and provide staff with training curricula with required timetables to complete, while others rely on staff to self-guide the training they need. We found many agencies contract out specific training or provide learning opportunities to staff on their intranet sites via e-learning tools. Most subject-matter experts said that misconduct training is likely more effective when delivered in-person, due to the broad range of issues related to misconduct.

OPM officials told us that supervisors and managers are responsible for observing and enforcing applicable laws in the federal workplace. OPM officials also indicated that training, resource allocation, skills, and knowledge all have a bearing on the administration of the disciplinary process. According to OPM, good communication and partnerships are also critical to processing a solid, sustainable response related to misconduct. OPM guidelines require that agencies provide training when employees make critical career transitions, for instance from nonsupervisory to manager or from manager to executive. Further, OPM's Supervisory and Managerial Curriculum Framework highlights human resources technical areas and leadership competencies necessary for success. The curriculum framework includes employee and labor relations with supporting learning objectives.

OPM has specific regulatory requirements for training and development of supervisors, managers, and executives under 5 CFR § 412.202, including to

- provide training within 1 year of an employee's initial appointment to a supervisory position and follow up periodically, but at least once every 3 years, by providing each supervisor and manager additional training on the use of appropriate actions, options, and strategies:
- to mentor employees;
- improve employee performance and productivity;
- conduct employee performance appraisals in accordance with agency appraisal systems; and
- identify and assist employees with unacceptable performance.[60]

According to 5 U.S.C. § 4103, it is the responsibility of each agency to train its employees. According to OPM officials, it is not responsible under the CSRA for providing training for the federal workforce. However, while agencies are accountable for providing required training for their

[60] 5 CFR 412.202(b)(1)-(4).

supervisors, OPM has a key role in ensuring the training meets the government-wide needs of supervisors. By taking steps to help agencies improve the training they provide supervisors and managers on addressing misconduct, OPM could help those managers ensure they have the knowledge and skills to effectively deal with misconduct in the workplace. For example, OPM could consider the feasibility of developing more in-person training modules designed to provide interactive or role play scenarios around addressing employee misconduct. Furthermore, subject-matter experts said if an agency is not training new supervisors to equip them with the appropriate skills to address misconduct, there may be inconsistencies in how an agency handles misconduct across the agency. Without sufficient training, supervisors and managers may not be addressing misconduct appropriately, if at all.

Internal Collaboration Can Help Agency Components Better Communicate about Misconduct Cases

Many of the subject-matter experts we interviewed said that it is important that the primary stakeholders—first-level supervisors and managers and human resources and general counsel offices—collaborate on the agency's approach to dealing with misconduct.

We found agencies vary in how collaboration takes place. For example, some subject-matter experts and CHCOs told us that an agency may choose to handle a case by having their human resources staff and management work closely together. The subject-matter experts we interviewed said this collaboration can sometimes include general counsel staff, if necessary. For example, at EPA, the human resources office collaborates with the office of general counsel and the agency's Office of Inspector General Office of Investigations (OI). EPA officials told us that their agency's human resources office, OI, general counsel, and labor relations meet bi-weekly to discuss ongoing misconduct investigations to provide a report of investigations to EPA's senior management on the facts surrounding allegations of employee misconduct.

According to EPA, OI also provides real-time notification whenever OI receives information concerning serious misconduct, before the investigation is completed, so EPA management can take appropriate immediate mitigating steps, should it be necessary. However, OI does not have a role in determining the type of discipline, if any, to be imposed upon the employee, nor does OI have any role in helping to prevent misconduct in EPA's workplace. We did not obtain data to verify that this process has been successful, but agree that enhanced communication among key stakeholders is important to addressing misconduct.

An Agency's Culture and Mission Can Impact How Agencies Approach and Respond to Misconduct

Most of our subject-matter experts told us an agency's culture and the nature of its work play a significant role in how the agency addresses employee misconduct. For example, several subject-matter experts told us law enforcement and defense-related agencies or other particular jobs where injuries may occur or lives may be at risk often have significantly less tolerance for employee misconduct than other agencies. In addition, OPM officials said that according to MSPB past studies, if an agency views federal employee due process procedural rights as burdensome and restrictive, this may discourage supervisors from addressing misconduct as it occurs. An MSPB report addressed concerns that the culture in many federal agencies prevents them from effectively dealing with problem employees.[61] Many of the subject-matter experts we interviewed indicated that if an agency's culture is risk averse, it may be less aggressive in pursuing adverse actions, and instead either ignore misconduct or reassign an employee without holding him or her accountable for the misconduct.

---

[61] Merit Systems Protection Board, *The Changing Federal Workplace: Employee Perspectives* (Washington, D.C.: March 1998).

CONCLUSION

The process for dismissing an employee for misconduct can be complex and lengthy. However, many of these process challenges can be avoided or mitigated with effective performance management. Supervisors who take performance management seriously and have the necessary training and support to address misconduct can help employees either change their conduct or be subject to removal from the federal workforce.

OPM has a role in ensuring that agencies have the tools and guidance they need to effectively address misconduct and maximize the productivity of their workforces. Though OPM already provides a variety of tools, guidance, and training to help agencies address issues related to misconduct, we found opportunities to do more to identify the nature of employee misconduct, improve training tools for managers, and make tools and guidance available for agencies when and where they need it.

RECOMMENDATIONS FOR EXECUTIVE ACTION:

We are making the following three recommendations to the Director of OPM:

- The Director of OPM, after consultation with the CHCO Council, should explore the feasibility of improving the quality of data on employee misconduct by providing additional guidance to agencies on how to record instances of misconduct in OPM's databases. (Recommendation 1)
- The Director of OPM, after consultation with the CHCO Council, should broadly disseminate to agencies the promising practices and lessons learned, such as those described in this chapter, as well as work with agencies through such vehicles as the CHCO Council, to identify any additional practices. (Recommendation 2)

- The Director of OPM, after consultation with the CHCO Council, should provide guidance to agencies to enhance the training received by managers/supervisors and human capital staff to ensure that they have the guidance and technical assistance they need to effectively address misconduct and maximize the productivity of their workforces. (Recommendation 3)

AGENCY COMMENTS AND OUR EVALUATIONS

We provided a draft of this product to the Acting Chairman of MSPB and Acting Director of OPM for comment. The Acting Chairman of MSPB provided technical comments on the draft. We incorporated these comments, as appropriate. MSPB did not comment on the recommendations.

OPM's Associate Director for Employee Services provided written comments on the draft, and these comments are reproduced in appendix III. In its comments, OPM noted that while we had made many of the changes OPM suggested, the changes still did not reflect all of OPM's feedback, and also contained what it believed to be inaccurate information and incomplete representations of OPM's views. To the contrary, we maintain that our report contains accurate factual information and represents the views of OPM that we collected through reviewing documents, interviewing OPM officials, and incorporating OPM's written feedback. Of our three recommendations, OPM partially concurred with two recommendations, and did not concur with one recommendation. For those recommendations OPM partially concurred with, OPM described the steps it planned to take to implement them. We stand by our recommendations which we maintain would give OPM and Congress better visibility over the extent and nature of employee misconduct in the federal government, as well as help strengthen agencies' capacity to address misconduct.

With respect to OPM's overall comments, OPM noted that Chapter 75 is a set of procedural requirements that must be met when certain actions are contemplated that would impact an employee's pay, specifying that it was never intended to encompass or catalogue all forms of action an agency could take to address misconduct. On this issue, we agree with OPM on the purpose of Chapter 75 and noted as much in our description of the statutorily established guidelines and procedures throughout this chapter. OPM also noted that there is no general statutory definition of misconduct, and that managers need maximum flexibility to pursue adverse actions, whether the underlying impetus is a conduct issue, a failure to perform, or any other reasons related to federal employment. We also agree with OPM on this point, as our report makes clear that, in certain cases, employee performance and misconduct can overlap, conflating the two issues.

As indicated in this chapter, OPM believes a table of penalties creates additional conditions and restrictions on an agency's ability to address misconduct and does not improve the agency's ability to address misconduct effectively. Accordingly, OPM does not require or encourage agencies to adopt tables of penalties. Our report recognizes both the pros and cons of an agency having a table of penalties and the circumstances under which they could be effective. However, we believe the use of a table of penalties ensures the appropriateness and consistency of a penalty in relation to the charge. It also ensures merit system principles guide the process by providing penalty transparency, reducing arbitrary or capricious penalties, and serve as a guide for managers and supervisors who deal with these issues.

With respect to our recommendations, OPM did not concur with our first recommendation to explore the feasibility of improving the quality of data on employee misconduct by providing additional guidance to agencies on how to record instances of misconduct in OPM's databases. Specifically, OPM noted that the OPM Guide to Processing Personnel Actions is a thorough resource that has been and continues to be successfully relied upon by agencies to document adverse actions as expressly defined in Chapter 75. We acknowledge OPM's view that NOA codes were never intended or designed to allow reporting of adverse actions down to the degree of a particular kind of misconduct involved, but we maintain that our recommendation would increase confidence in the data on misconduct and make it more useful to OPM and agencies. Further, OPM's non-concurrence with this recommendation seems inconsistent with the Administration's own initiatives, including the May 2018 *Executive Order Promoting Accountability and Streamlining Removal Procedures Consistent with Merit System Principles*, which was released after OPM commented on our draft report.[62] Specifically, the Executive Order requires all federal agencies, beginning in FY18, and for each fiscal year thereafter, to provide a report to the OPM Director containing detailed data about how it addressed issues of misconduct. For example, agencies will need to report out on

I. the number of civilian employees in a probationary period or otherwise employed for a specific term who were removed by the agency;
II. the number of adverse personnel actions taken against civilian employees by the agency, broken down by type of adverse personnel action, including reduction in grade or pay (or equivalent), suspension, and removal; and
III. the number of decisions on proposed removals by the agency taken under chapter 75 of title 5, United States Code, not issued within 15 business days of the end of the employee reply period.

We maintain that enhanced data on the extent and nature of misconduct will help strengthen OPM and congressional oversight and better position agencies to address misconduct through management training and other approaches.

OPM partially concurred with our second recommendation to broadly disseminate to agencies the promising practices and lessons learned, such as those described in this chapter, as well as work with agencies through

---

[62] *Executive Order 13839: Promoting Accountability and Streamlining Removal Procedures Consistent with Merit System Principles* (Washington, D.C.: May 25, 2018).

such vehicles as the CHCO Council, to identify any additional practices to help agencies better address employee misconduct. Indeed, the *President's Management Agenda* (PMA) for 2018 states, "Aligning and managing the Federal workforce of the 21st Century means spreading effective practices among human resources specialists."[63] In response to this recommendation, OPM noted that some of the key practices and lessons discussed in this chapter are already part of OPM's comprehensive accountability toolkit in addressing employee misconduct across the federal government and are frequently communicated through on-going educational outreach to federal agencies and available on OPM's website. Specifically, OPM said it will decide which appropriate measures it should take to obtain examples of practices agencies believe are promising and will broadly disseminate any of these practices and lessons learned as identified by OPM. We acknowledge OPM's existing efforts to develop and disseminate promising practices and lessons learned, and also maintain that OPM should also be open to considering additional practices from other sources.

OPM also partially concurred with our third recommendation to provide guidance to agencies to enhance the training received by managers/supervisors and human capital staff to ensure that they have the guidance and technical assistance they need to effectively address misconduct and maximize the productivity of their workforces. In its response, OPM said it will continue to play its statutory role under 5 U.S.C. Chapter 41 and will support agencies on a cross-agency priority goal, which it believes could be read to encompass training, pursuant to the PMA, for example by providing guidance to agencies on training requirements for managers, supervisors and human resources staff. However, OPM notes that it is not responsible under current statute for providing training to the federal workforce. As stated in the report, while agencies are accountable for providing required training for their supervisors, OPM has a key role in ensuring the training meets the needs of supervisors. Further, OPM's position on this recommendation seems inconsistent with the Administration's own initiatives, including the May

---

[63] *The President's Management Agenda 2018: Modernizing Government for the 21st Century* (Washington, D.C.: March 20, 2018).

2018 Executive Order which states that "the OPM Director and the Chief Human Capital Officers Council shall undertake a Government-wide initiative to educate Federal supervisors about holding employees accountable for unacceptable performance or misconduct under those rules," following any final rules issued pursuant to parameters set in the Order. Indeed, the PMA states, "In order to best leverage the workforce to achieve our mission efficiently and effectively, Government needs to remove employees with the worst performance and conduct violations." By taking steps to help agencies improve the training they provide supervisors and managers on addressing misconduct, OPM could help those managers ensure they have the knowledge and skills to effectively deal with misconduct in the workplace.

OPM also provided technical comments, which we have incorporated as appropriate.

As agreed with your office, unless you publicly announce the contents of this chapter earlier, we plan no further distribution until 30 days from the report date. At that time, we will send copies to the Director of the Office of Personnel Management, the Chairman of the Merit Systems Protection Board, as well as to the appropriate congressional committees and other interested parties.

Robert Goldenkoff
Director
Strategic Issues

APPENDIX I: OBJECTIVES, SCOPE, AND METHODOLOGY

You asked us to examine the process for addressing misconduct and to identify any challenges in removing employees for misconduct. Our objectives were to (1) describe the process that agencies are generally required to follow in responding to employee misconduct in the federal service; (2) identify alternative approaches to the formal legal process that agencies can use to respond to misconduct, and assess what factors affect agencies' responses; (3) describe trends in removals and other adverse actions resulting from misconduct; and (4) identify key steps agencies can take to help them better prevent and address misconduct.

To describe the process that most agencies are generally required to follow in responding to employee misconduct in the federal service, we reviewed relevant sections of Title 5 Chapter 75 of the U.S.C. (herein Chapter 75) which contains the statutory process for formally disciplining employees for misconduct and performance.[64] We also reviewed the Civil Service Reform Act and OPM regulations to describe and determine the authority agencies have to address employee misconduct in the federal service, including formal procedural and employee appeal rights.[65] Additionally, we reviewed 5.U.S.C. §§ 7701 and 7702 (herein Chapter 77), which contains the statutory process for employee appeals with the Merit Systems Protection Board (MSPB) subsequent appeals to the Equal Employment Opportunity Commission (EEOC).[66] To contextualize our review of Chapters 75 and 77, we interviewed current and former practitioners, subject-matter experts, and academics. We selected our subject-matter experts based on our guidance for selecting experts, including their depth of experience on this issue and their present and past employment history, the subject-matter experts' practical experience in applying and practicing administrative law related to employee misconduct, and academicians who have conducted research on employee misconduct.

To identify the time it takes to resolve appeals by employees of adverse actions, we analyzed MSPB appeals data. Specifically, we counted the appeals for each type of adverse action from 2006-2016, by year and in aggregate, calculated the average number of days that it took the MSPB to resolve cases for initial appeals and for petitions for review (PFRs), and counted the outcomes of the cases from 2006-2016, by year and in aggregate (e.g., out of the total number of cases, how many were reversed, upheld, mitigated, or settled). For the purpose of our analysis, we used the

---

[64] 5 U.S.C. §§ 7501-7514.

[65] OPM regulations implementing Chapter 75 are found in 5 C.F.R. subpart D of part 752.

[66] 5 U.S.C. § 7701(a)-(b); 5 U.S.C. § 7702(b).

following nature of action (NOA) code categories in OPM's EHRI database: (1) codes directly attributed to misconduct; and (2) codes that indicate a mix of misconduct or poor performance. Based on data limitations in both databases, we did not make any evaluative assessments from our data analysis.

To identify alternative approaches to the formal legal process, we reviewed documentation provided by the Merit System Protection Board (MSPB) on alternative discipline approaches used by agencies to address employee misconduct. We also reviewed OPM regulations and documents to determine the authority agencies have to address employee misconduct in the federal service, including formal procedural and employee appeal rights. We interviewed current and former practitioners, subject-matter experts, and academics to identify alternative approaches that they were aware of or were commonly used at agencies to address employee misconduct.

To develop our list of alternative discipline approaches to addressing employee misconduct, we conducted a literature review and reviewed reports and documents to identify alternative discipline approaches commonly used to address employee misconduct in the federal sector. After compiling our non-exhaustive list of alternative approaches, we contacted our previously interviewed subject-matter experts and asked them to provide their final thoughts or suggestions to our alternative discipline approaches. We included those additional approaches to the list. We interviewed human capital experts from academia, unions, and former and current human resources practitioners. We also interviewed a panel of CHCOs to gain insight into the agency perspective on addressing employee misconduct. To identify CHCO members, we asked the Director of the CHCO council to select CHCOs that have knowledge and experience in addressing employee misconduct. Agency size and mission were also considered as part of the selection process to gain a range of perspectives. Our panel of CHCOs was from the Departments of Commerce, Defense, and Housing and Urban Development, the National Science Foundation, and the Nuclear Regulatory Commission. We also reviewed prior work by

MSPB in developing our list of commonly used alternative discipline approaches to employee misconduct in the federal sector.

To describe and assess the factors that can affect an agency's response to employee misconduct, we interviewed:

- OPM officials and representatives from Employee Services, Human Resources Solutions, Planning and Policy Analysis, and the Office of the Chief Information Officer
- MSPB officials from the Office of the acting Chairman & Vice Chairman, Office of Information Resources Management, and the Office of Policy & Evaluation;
- Panel of Chief Human Capital Officers (CHCO)
- National Treasury Employees Union officials;
- American Federation of Government Employees officials;
- Federal Managers Association officials;
- Individual members of the Federal Employees Lawyers Group;
- Partnership for Public Service officials;
- Senior Executives Association officials; and
- Selected individuals with expertise in human capital management, specifically focused on employee misconduct, from academia and the private sector.

We selected our list of interviewees based on GAO's guidance for selecting experts, the interviewees' practical experience in applying and practicing administrative law, and for academics in their specific areas of research. To assess the factors that agencies use to deal with employee misconduct, we analyzed the interviewee responses and identified key themes that were common throughout our interviews and, we counted the frequency of those key themes.

To describe the trends in removals and adverse actions resulting from misconduct at Chief Financial Officer (CFO) Act agencies, we analyzed OPM's Enterprise Human Resource Integration (EHRI) data from fiscal years 2006 to 2016.[67] In OPM's EHRI database, personnel actions, such as separations, demotions, and suspensions, are assigned nature of action (NOA) and legal authority codes that describe the action and the legal or regulatory authority for the action. Some of these codes co-mingled performance and misconduct as causes for the personnel action. In these cases, we counted the action as misconduct-related to capture the potential magnitude of personnel actions taken for misconduct. Thus, some cases may instead be exclusively related to poor performance or reflect both issues. We reviewed OPM's "The Guide to Processing Personnel Actions" to determine which NOA/legal authority combinations could cover misconduct-related actions. We worked with OPM officials to determine the appropriate NOA codes and legal authorities that indicate misconduct.

Specifically, we analyzed OPM's EHRI data to count the adverse actions that agencies take against employees, government-wide.[68] For this draft report, the term "adverse action" refers to personnel actions listed under 5 U.S.C. chapter 75 which includes demotions in grade or pay, suspensions, and removals.[69]

To determine the trends in the number of adverse actions at agencies associated with probationary employees, we identified the number of removals at these agencies. While direct comparisons to the total employee population are not appropriate, we identified the number of probationary

---

[67] The 24 CFO Act agencies are listed at 31 U.S.C. § 901(b) and include: U.S. Departments of Agriculture, Commerce, Defense, Education, Energy, Health and Human Services, Homeland Security, Housing and Urban Development, the Interior, Justice, Labor, Transportation, the Treasury, Veterans Affairs, and State, as well as the U.S. Agency for International Development, Environmental Protection Agency, General Services Administration, National Aeronautics and Space Administration, National Science Foundation, Nuclear Regulatory Commission, Office of Personnel Management, Small Business Administration, and the Social Security Administration. These agencies account for a very high proportion of the total federal labor force.

[68] For reporting purposes, we only provide data on the number of adverse actions rather than use the number of employees subject to an adverse action, because one employee may be subject to more than one adverse action. This can be a result of progressive discipline or could indicate issues related to data reliability. As part of our analysis, we identified all employees subject to each type of adverse action (removal, suspension, and demotion) and quantified the number of similar adverse actions taken against the same person.

[69] *5 U.S.C. §§ 7502 and 7512.*

employees to provide some relative statistics.[70] To show trends in misconduct removals associated with probationary employees, we identified the number of adverse actions taken against probationary employees (overall and by type of action).

To determine the trends in employee appeals to MSPB, we analyzed MSPB's appeals data, on adverse actions taken under Chapter 75 from fiscal years 2006 to 2016. To understand what types of adverse actions are driving appeals to MSPB, we calculated the underlying adverse action for each case by fiscal year of appeal filing.[71] To determine how appeals were resolved by MSPB, we identified the number of appeals that were settled, mitigated, dismissed, reversed, affirmed, or otherwise resolved. We calculated overall trends by fiscal year as well as trends by fiscal year for each type of underlying adverse action. To determine how long the appeals process takes, we calculated the mean time for resolution along with other statistics (minimum, maximum, 25th percentile, 75th percentile) for different types of adverse action. Additionally, because appellants can file a Petition for Review (PFR) to the larger MSPB body, we looked at the time for the initial appeal, the PFR, and the total time from filing through final decision.

To assess the reliability of both EHRI and MSPB data, we reviewed past GAO data reliability assessments, interviewed relevant agency officials, and conducted electronic testing to evaluate the accuracy and completeness of the data used in our analyses. We determined the data used in this chapter to be sufficiently reliable for our purposes, subject to the constraints identified in our report.

---

[70] Due to fluctuations in employment, it is GAO's policy to compute a 2-year average level of employment.

[71] Some people file several appeals related to the same adverse action. To avoid over-counting the underlying adverse action, we deleted all but the first case (identified as having the same case family key and case key) in this table. We do not do this for the other tables, because outcomes may vary for each appeal.

To identify and provide key promising practices and lessons learned at agencies from encountering and responding to employee misconduct, we conducted a literature review to identify practices and lessons learned associated with employee misconduct in the federal sector. We interviewed officials from OPM, MSPB, the Equal Employment Opportunity Commission (EEOC), and the Office of Special Counsel (OSC), to obtain their perspectives on responding to employee misconduct through alternative approaches. We interviewed officials from the Environmental Protection Agency (EPA) to obtain their perspectives on recent efforts to better coordinate with their Inspector General to address cases of employee misconduct.[72] We also obtained the perspectives of a panel of CHCOs from selected agencies as well as former human capital practitioners and other subject-matter experts with extensive experience working on employee misconduct issues.

APPENDIX II: DOUGLAS FACTORS – 12 CRITERIA DEVELOPED BY THE MSPB TO GUIDE AGENCY DECISIONS ON EMPLOYEE MISCONDUCT

The Merit Systems Protection Board in its landmark decision, *Douglas vs. Veterans Administration*, 5 M.S.P.R. 280 (1981), established non-exclusive criteria that supervisors must consider, as appropriate, in determining an appropriate penalty to impose for an act of employee misconduct ("The Douglas Factors"). The following relevant factors must be considered in determining the severity of the discipline:

1) The nature and seriousness of the offense, and its relation to the employee's duties, position, and responsibilities, including whether the offense was intentional or technical or inadvertent, or was committed maliciously or for gain, or was frequently repeated;

[72] We met with EPA officials because of the agency's relatively unique process for addressing employee misconduct.

2) the employee's job level and type of employment, including supervisory or fiduciary role, contacts with the public, and prominence of the position;
3) the employee's past disciplinary record;
4) the employee's past work record, including length of service, performance on the job, ability to get along with fellow workers, and dependability;
5) the effect of the offense upon the employee's ability to perform at a satisfactory level and its effect upon supervisors' confidence in the employee's work ability to perform assigned duties;
6) consistency of the penalty with those imposed upon other employees for the same or similar offenses;
7) consistency of the penalty with any applicable agency table of penalties;
8) the notoriety of the offense or its impact upon the reputation of the agency;
9) the clarity with which the employee was on notice of any rules that were violated in committing the offense, or had been warned about the conduct in question;
10) the potential for the employee's rehabilitation;
11) mitigating circumstances surrounding the offense such as unusual job tensions, personality problems, mental impairment, harassment, or bad faith, malice or provocation on the part of others involved in the matter; and
12) the adequacy and effectiveness of alternative sanctions to deter such conduct in the future by the employee or others. The list was not intended to be exhaustive.

APPENDIX III: COMMENTS FROM THE OFFICE OF PERSONNEL MANAGEMENT

UNITED STATES OFFICE OF PERSONNEL MANAGEMENT
Washington, DC 20415

Employee Services

MAY 30 2018

Mr. Robert Goldenkoff
Director, Strategic Issues
U.S. Government Accountability Office
441 G Street, NW
Washington, DC 20548

Dear Mr. Goldenkoff:

Thank you for providing us the opportunity to respond to the U.S. Government Accountability Office (GAO) draft report, *Federal Employee Misconduct: Actions Needed to Ensure Agencies Have Tools to Effectively Address Misconduct,* GAO-18-48, GAO job code number 101023.

We acknowledge that GAO made many of the changes we suggested, and we appreciate that effort. In significant areas, however, the report does not reflect the full import of our feedback and retains both substantive inaccuracies and what we believe to be inaccurate or incomplete representations of OPM's views. We have attached technical comments to this letter. However, we highlight below substantive concerns regarding the report.

More generally, we believe it is important to note that the recommendations GAO has made appear to proceed, in part, from misunderstandings about the nature of discipline generally, and the purpose of Chapter 75 more specifically. The purpose of Chapter 75 was not to describe modes of "formal discipline." The authority to undertake discipline predates the enactment of Chapter 75, which was intended to establish certain statutory limitations on the plenary authority to impose discipline, not to create the authority in the first instance. Chapter 75 established procedures that must be followed when an agency wishes to take certain actions against an employee that would have an impact upon his or her pay, *i.e.*, "adverse actions." Adverse actions can be taken for any reason that would promote the efficiency of the service. Such reasons could include, but are not limited to, misconduct. For example, an adverse action may be used to remove an employee who is not performing at the fully successful level or to remove or demote an employee who has lost eligibility for access to classified information in a position that requires such eligibility, or to effectuate an administrative furlough.

There is no general statutory definition of misconduct, and Congress has not asked or authorized OPM to define or list forms of misconduct. By establishing procedures for effectuating adverse actions, and supplying a general substantive standard for doing so (the efficiency of the service), Congress provided managers with maximum flexibility to pursue adverse actions, whether the underlying impetus is a conduct issue, a failure to perform, or any other reasons related to Federal employment. GAO's focus on capturing and tabulating forms of misconduct (and misconduct alone) would undermine this flexibility and the discretion Congress continued to allow agency heads by forcing managers to cabin adverse actions in terms of specific definitions that may not capture the full scope of the considerations involved. OPM does not fully understand why GAO believes this is necessary or desirable for successful workforce management or why it is seemingly troubling to GAO that some agencies take adverse actions for reasons described as both disciplinary and performance-related. In some situations, the two will overlap. Sleeping on the job, for example, is both a failure to abide by norms of conduct an agency would expect of an employee on duty and a failure to perform. OPM does not see that precisely defining "sleeping on the job" as an official form of misconduct does anything to advance an agency's ability to effectively address the behavior and promote the efficiency of the service. Further, OPM does not understand how collecting and aggregating granular data on instances of low level offenses across a large Federal workforce would be instructive to overall enterprise workforce management matters. For example, we would question the value of tabulating how many employees across the 2 million person enterprise may have received verbal counseling for failing to follow a supervisor's instruction on accomplishing a work task, or for wearing attire that is considered outside of norms for the local workplace.

As we noted above, Chapter 75 is a set of procedural requirements that must be met when certain actions are contemplated. It was never intended to encompass or catalogue *all* forms of action an agency could take to address misconduct, and, in OPM's experience, agencies readily understand that other forms of actions, such as counseling or reprimands, are available to them in dealing with employees who are not comporting themselves appropriately. These other forms of disciplinary actions are not subject to OPM regulation. For that reason, it is likely that the number of employees who were subjected to some form of discipline, as GAO uses that term, is much larger than the number reflected in the draft report as Chapter 75 actions.

Regarding tables of penalties, the draft refers on the first page to tables of penalties as an example of "key lessons learned" and elsewhere in the draft as a "key promising practice" that can "help agencies better prevent and respond to misconduct." The report does not, however, explain how having a table of penalties will help an agency to prevent misconduct or respond to it. To the extent GAO is assuming that the existence of a table of penalties will serve as a warning to employees or compel supervisors to carry out more disciplinary actions for the conduct identified in the table, it is equally likely that the table of penalties will de-emphasize constructive early intervention in favor of a more punitive approach that focuses only on the offenses the table covers. It may also be read or understood to induce or worse, require, managers in some cases to impose a lesser penalty where a greater penalty is warranted.

Although the draft report references some of OPM's concerns about tables of penalties, neither the summary observation on the first page of the report, nor the overall report, contains a serious discussion of the disadvantages of a table of penalties, which we believe are important considerations in assessing their value. In addition to concerns referenced above, OPM has emphasized, for example, that it is vital for effective workforce management consistent with the CSRA and merit system principles that supervisors use independent judgment, take appropriate steps in gathering facts, and conduct a thorough analysis to decide the appropriate penalty in individual cases. Further, we explained that applying a label to an act of misconduct (which is a practice that is encouraged by a table of penalties) can lead – and has led - to reversals of adverse actions by the MSPB on extremely technical grounds, (e.g., by finding that a charge of "theft" of an inmate's petty cash by a corrections officer could not be sustained where the agency could not prove that the officer intended to keep the cash permanently). We also noted that the MSPB will sometimes invoke an agency's own table of penalties against it to support the MSPB's decision to mitigate a penalty, (from, for example, a removal to a suspension), while treating a table of penalties as merely advisory if it supports the agency's decision. An agency might decide to define inappropriate conduct in its policies rather than in a table of penalties to avoid these pitfalls. Accordingly, OPM believes it would be preferable to include a more balanced discussion of tables of penalties in the GAO report.

Responses to your recommendations are provided below.

Recommendation #1:

The Director of OPM, after consultation with the CHCO Council, should explore the feasibility of improving the quality of data on employee misconduct by providing additional guidance to agencies on how to record instances of misconduct in OPM's databases.

Management Response:

We do not concur. The draft report does not identify any basis or clear rationale for how collecting and aggregating granular data on specific instances of misconduct across a large Federal workforce would be instructive to overall enterprise workforce management matters. The OPM Guide to Processing Personnel Actions is a thorough resource that has been and continues to be successfully relied upon by agencies to document adverse actions as expressly defined in Chapter 75.

GAO articulated a concern about OPM's inability to accurately identify misconduct cases in the Enterprise Human Resources Integration (EHRI) system because of nature of action (NOA)/legal authority codes that are a combination of both conduct and performance. We acknowledge that there is an instance where a particular NOA could be used for different reasons, and accordingly, our data collection does not always differentiate between actions strictly for performance, actions for misconduct, or both. While we note that agencies use this code infrequently, and, therefore, we believe the impact on the overall data is negligible, we plan to review this NOA for possible changes to address GAO's concern.

The draft states, that "[w]ithout maintaining comprehensive data regarding the extent and nature of misconduct in the Federal government, OPM risks missing opportunities to provide agencies with guidance and other tools, such as targeted training to help agencies better address cases of misconduct." Further, the draft asserts that "these data could be improved to provide OPM with better information to help agencies address misconduct." NOA codes were never intended or designed to allow reporting of adverse actions to the degree of specificity envisioned in the GAO report. In particular, they do not break down adverse actions by the particular kind of misconduct involved, and in light of the enormous array and complexity of types of misconduct that could potentially be encountered across a large workforce, it would seem neither practicable nor useful to attempt to precisely define and count each type of misconduct, let alone the many permutations that could occur when a particular instance of misconduct is multifaceted. The development of NOA codes proposed by GAO would require not only extensive changes in recordkeeping and reporting requirements throughout the Government, but would also add enormous complexity, room for error, and burden as agency HR staff would have to sort through and select from potentially hundreds of codes or combinations thereof.

The draft report acknowledges that EHRI does not capture the various types of informal actions managers may take that do not generate an SF-50, such as counseling letters or letters of reprimand, which may correct an issue before it escalates into a formal adverse action. Although there may be some value in broadly enumerating the numbers of certain types of disciplinary actions, we do not believe these types of actions warrant the processing of a personnel action for every instance for purposes of tracking in EHRI. GAO also appears to suggest that OPM needs to collect data on types of misconduct resulting in discipline that do not impact pay (and consequently does not require that an SF-50/personnel action be processed) such as letters of reprimand or counseling, which OPM does not regulate. Although a particular agency may potentially be interested in quantifying such types of activity, we question the substantive value in collecting such granular data at an enterprise level. Furthermore, by establishing additional administrative requirements for supervisors to report the entirety of instances of misconduct (which presumably would include lower-level offenses such as verbal counseling when an employee fails to follow instructions related to accomplishing a task, or for wearing attire that is considered outside of norms for the local workplace) this could have the effect of discouraging supervisors from taking action in the first instance to attempt to correct behavior before it worsens.

Another important factor is that the recommendation appears to be based on the incorrect premise that OPM is responsible for developing and delivering supervisory training. The draft states that "because of weaknesses in OPM's data on employee misconduct, which is provided by the agencies, OPM is unable to accurately target supervisory training to address misconduct, and decision-makers do not know the full extent or nature of this misconduct." As discussed above, the draft also states that, without maintaining comprehensive data on misconduct, OPM may not be able to provide agencies tools such as training to aid in addressing misconduct.

As discussed more fully under Recommendation 3, below, each agency, and <u>not</u> OPM, has responsibility for developing and delivering training to its supervisors. OPM provides guidance and technical assistance which agencies are expected to apply in developing their own training. Given the fact that longstanding, comprehensive guidance exists for agencies to use in successfully coding misconduct actions, and the questionable value of significantly expanding data collection and reporting to the extent the GAO proposal suggests, OPM respectfully requests this recommendation be deleted from the final report.

Finally, the strategic management of human capital is primarily an OPM function. See, e.g., 5 U.S.C. § 1103(a)(5) and (7) and (c). We note that the statute creating the CHCO Council stated that the role of the CHCO was to "advise and assist *the head of the agency* and other agency officials in carrying out the agency's responsibilities for selecting, developing, training, and managing a high-quality, productive workforce in accordance with merit system principles," and to "implement the rules and regulations of the President and the Office of Personnel Management and the laws governing the civil service within the agency." 5 U.S.C. § 1401(1), (2). (Emphasis supplied). And, according to the CHCO Act, the function of the CHCO Council meetings was "to advise and coordinate *the activities of the agencies of its members* on such matters as modernization of human resources systems, improved quality of human resources information, and legislation affecting human resources operations and organizations." Pub. L. 107-296, § 1303(b). (Emphasis supplied). Although the Director is, of course, free to consult with the CHCO Council as he chooses -- and may well wish to consult with the CHCO Council on these matters -- it is not appropriate for GAO to attempt to impose such extra-statutory requirements on the Director through recommendations in connection with an engagement.

Recommendation #2:

The Director of OPM, after consultation with the CHCO Council, should broadly disseminate to agencies the promising practices and lessons learned, such as those described in this report, as well as work with agencies through such vehicles as the CHCO Council, to identify any additional practices.

Management Response:

We partially concur only to the extent that OPM agrees to continue its ongoing efforts through the accountability toolkit and related outreach. OPM already has a mandate under the Civil Service Reform Act to identify policies which may be implemented to promote an efficient civil service and a systematic application of merit system principles, and will continue to fulfill its responsibilities under the Act. Although OPM may choose to solicit (and in fact has solicited) from agencies examples of practices they believe are promising and to broadly disseminate to agencies any promising practices and lessons learned as identified by OPM, it must be free to do so on its own terms consistent with its own authority. OPM is not willing to abdicate its authority to determine what constitutes a promising practice and how it communicates those practices. Accordingly OPM cannot concur in this recommendation to the extent it commits OPM to broadly disseminate "promising practices" as identified by GAO. We note that some of the key practices and lessons discussed in the draft are already part of

OPM's comprehensive accountability toolkit in addressing employee misconduct across the Federal government and are frequently communicated through on-going educational outreach to Federal agencies and available on OPM's website. OPM provided this information to GAO on January 12, 2017. OPM does not agree that all the practices outlined in GAO's report are promising practices which should be disseminated to agencies. For example, GAO's report suggests that tables of penalties are "key lessons learned" and "key promising practices." OPM has articulated to GAO the various challenges and pitfalls associated with application of tables of penalties, and we believe that an agency should take careful heed of these admonitions in considering whether a table of penalties actually helps them prevent or respond more effectively to employee misconduct or impedes their efforts. For that reason, we believe the characterization of tables of penalties as a "key lesson learned" and a "key promising practice" is problematic in that it seems to encourage adoption without fully articulating the downsides. For the foregoing reasons, OPM must partially concur. OPM respectfully requests the recommendations regarding table of penalties as a key lesson learned and key promising practice be deleted from the final report or OPM's views be entirely included to ensure our concerns are represented. As for the role of the CHCO Council, please see the discussion at the end of our response to Recommendation # 1.

Recommendation #3:

The Director of OPM, after consultation with the CHCO Council, should provide guidance to agencies to enhance the training received by managers /supervisors and human capital staff to ensure that they have the guidance and technical assistance they need to effectively address misconduct and maximize the productivity of their workforces.

Management Response:

We partially concur only to the extent that OPM will continue to play its statutory role under 5 U.S.C. chapter 41 and will support agencies on a cross agency priority goal, which could be read to encompass training, pursuant to the President's Management Agenda. This recommendation and some of the rationale provided in GAO's report are not consistent with the clearly defined responsibilities of OPM and Federal agencies with regard to training and how it is delivered and funded. As we previously explained, OPM is not responsible under current statute for providing training to the Federal workforce. In accordance with its general oversight and policy responsibilities for the Federal personnel system, and in accordance with 5 U.S.C. chapter 41, OPM gives guidance to agencies on training requirements for managers, supervisors and human resources staff. Agencies are responsible for applying this guidance when developing and delivering their own training. OPM otherwise provides technical assistance and other support to agencies, as needed. To the extent OPM is involved in delivering training, OPM does so through OPM's human capital consultants, Human Resources Solutions, on a fee-for-service basis. In this case, OPM provides a service akin to what a contractor might otherwise perform.

The President's Management Agenda (PMA) includes a cross agency priority goal regarding the "Workforce for the 21st Century." One sub-goal is to "Improve Employee Performance Management and Engagement." One of the key milestones under this subgoal is to ensure that managers are appropriately trained on performance management (including misconduct) and are provided with support to address performance and conduct issues. OPM will be supporting agencies in these efforts, consistent with the clearly defined responsibilities of OPM and Federal agencies with regard to training and how it is delivered and funded. OPM will also continue to provide agencies with guidance and technical assistance regarding misconduct matters. Therefore, OPM believes that actions taken in conjunction with chapter 41 and the PMA will appropriately address the concerns and issues raised by GAO regarding training received by managers/supervisors and human capital staff. As for the role of the CHCO Council, please see the discussion at the end of our response to Recommendation # 1.

Unless otherwise noted, the suggested revisions we recommend in the attached technical comments are meant to provide technical accuracy and correct statements of position that may be attributable to OPM.

I appreciate the opportunity to respond to this draft report. If you have any questions regarding our response, please contact Ms. Janet Barnes, Director, Internal Oversight and Compliance, (202) 606-3207 or Janet.Barnes@opm.gov.

Sincerely,

Mark D. Reinhold
Associate Director for Employee Services

INDEX

D

E

I

J

K

L

M

N

O

P

R

S

T

U

V

W